Springer Praxis Books

Space Exploration

The *Springer-Praxis Space Exploration* series covers all aspects of human and robotic exploration, in Earth orbit and on the Moon and planets. The books tell behind-the-scenes stories of early and modern missions, both crewed and uncrewed, and cover all aspects of the space programs run by both leading and emerging spacefaring nations.

* * *

The books in this series are well illustrated with color figures and photographs. They are written in a style that space enthusiasts and historians, readers of popular magazines such as *Spaceflight* and readers of *Popular Mechanics* and *New Scientist* will find accessible.

Pamela Elizabeth Clark

Extreme Exploration

Celebrating And Applying Lessons Learned From The Apollo Expeditions

Pamela Elizabeth Clark
Space Science Center Faculty and Star Theater Director
Morehead State University
Morehead, KY, USA

ISSN 2945-7475 ISSN 2945-7483 (electronic)
Springer Praxis Books
ISSN 2731-5401 ISSN 2731-541X (electronic)
Space Exploration
ISBN 978-3-032-01733-8 ISBN 978-3-032-01734-5 (eBook)
https://doi.org/10.1007/978-3-032-01734-5

© The Editor(s) (if applicable) and The Author(s), under exclusive license to Springer Nature Switzerland AG 2025

This work is subject to copyright. All rights are solely and exclusively licensed by the Publisher, whether the whole or part of the material is concerned, specifically the rights of translation, reprinting, reuse of illustrations, recitation, broadcasting, reproduction on microfilms or in any other physical way, and transmission or information storage and retrieval, electronic adaptation, computer software, or by similar or dissimilar methodology now known or hereafter developed.
The use of general descriptive names, registered names, trademarks, service marks, etc. in this publication does not imply, even in the absence of a specific statement, that such names are exempt from the relevant protective laws and regulations and therefore free for general use.
The publisher, the authors and the editors are safe to assume that the advice and information in this book are believed to be true and accurate at the date of publication. Neither the publisher nor the authors or the editors give a warranty, expressed or implied, with respect to the material contained herein or for any errors or omissions that may have been made. The publisher remains neutral with regard to jurisdictional claims in published maps and institutional affiliations.

This Springer imprint is published by the registered company Springer Nature Switzerland AG
The registered company address is: Gewerbestrasse 11, 6330 Cham, Switzerland

If disposing of this product, please recycle the paper.

This book is especially meant to honor the Apollo era astronauts, the most visible part of the Apollo effort, who were my heroes during childhood, as I ate, slept, and breathed the Apollo Program. I'm especially grateful for both President John F. Kennedy's and Werner von Braun's vision for space exploration that resulted in the creation and success of the Apollo Program. I met von Braun when I was barely a teenager during an activity, sponsored by the Girl Scouts that provided the model for future "Space Camps." That being said, this book is entirely dedicated to the hundreds of thousands of engineers, scientists, program managers, technicians, and political leaders who made possible the Apollo expeditions, one of the most extraordinary achievements of human history. Jim Lovell's famous comment on this effort that inspired me then continues to inspire me now and says it best: "The astronaut is only the most visible member of a very large team, and all of us, right down to the guys sweeping the floor are honored to be a part of it.... This is divine inspiration, folks. It's the best part of each one of us that anything is possible...".

Foreword

The Once and Future Moon: Once

Half a century after Apollo astronauts landed on the Moon, NASA is preparing to return people to our nearest neighbor in space. What can we learn from Apollo that will help guide us on this great enterprise? Pamela Clark's new book looks back to Apollo and ahead to Artemis to answer that question.

This book begins with the Apollo missions, including excerpts from interviews by some of the people involved: geologists Farouk El Baz, Lee Silver, and Jim Head as well as the geologist best equipped to advise on future exploration, Harrison Schmitt. El Baz worked on landing site selection, Silver on astronaut training, and Head on planning surface activities. Schmitt is the geologist-astronaut who flew on Apollo 17 and is still, over 50 years later, working to understand the results of his fieldwork. They saw what worked well and what did not, offering essential advice for future planning.

The book then delves into pre-Apollo exploration in search of useful analogs—Lewis and Clark in the western United States, Stanley in Africa, and perhaps most significantly the extraordinary Antarctic expeditions of Scott, Shackleton, and Amundsen. Apollo never had a mission like Scott's, thankfully, but it had a mission like Shackleton's (Apollo 13) and six more like Amundsen's, and looking into these analogs is illuminating. I have not seen this connection made so constructively before.

Robotic missions preceded Apollo and provided essential information for planning. Ranger spacecraft hit the Moon, taking high resolution pictures on the way down, giving a first close-up view of the lunar surface. Surveyor soft-landed to view and probe the surface. Lunar Orbiter viewed many potential landing sites at high resolution. Was the surface strong enough to support a landed spacecraft? Where would we find the least hazardous sites, the ones suitable for the first landings? Where were the most scientifically interesting sites for later missions?

Those early chapters set the stage for Apollo planning, and a great strength of this book is the detailed description of many different aspects of Apollo planning. How were astronauts chosen in the first place? How were they trained, not just on the hardware but in field geology to simulate what they might encounter on the Moon? How would they operate sophisticated tools and equipment while wearing bulky suits and gloves? How would they keep track of everything they had to do on the surface? (They wore small notebook-like checklists on their suit cuffs.) How would they navigate to the places where they were supposed to collect rock samples, and find their way home again?

All the tools had to be designed for operation in this harsh environment by people wearing thick gloves. They didn't always work well and some had to be redesigned between missions. Each crew left science instruments on the surface, and they had to be selected, designed, and then deployed on the Moon as quickly as possible in the limited time available. All these aspects of Apollo planning will have to be replicated as new missions are planned. We can see how some old problems will be overcome differently in future, such as linking surface experiments to a power source and communication unit. Apollo did it with cables, leading to the very unfortunate loss of a heat flow experiment on Apollo 16 when an astronaut's boot pulled a cable off its connection. In future, wireless communication and perhaps wireless charging technologies will be available, avoiding the need for a mass of cables, and these are being designed and tested now. But this sort of innovation will be needed for every aspect of the forthcoming Artemis missions.

After a summary of the Apollo missions which flew to the Moon in 1969–1972, and a brief look at three more which were cancelled, this book concludes with a thorough summary of these new challenges. Many of the people who did this work for Apollo are now lost to us, so there has been a huge loss of direct experience, of institutional memory. Luckily, much was recorded after the landings, such as the interviews mentioned earlier. But many things will have to be reinvented, and of course technology has changed so much that past experience has its limitations anyway. Also, with time to digest the experience of Apollo, new questions are being asked. For instance, prolonged activity on the surface makes the difficulties created by lunar dust more complex than during the brief Apollo missions. Work to mitigate this is being actively pursued today, and is well summarized here. This final chapter looks at many related issues and will be useful to anybody thinking about our future on the Moon. Pamela Clark has performed a valuable service by collecting all this material into a useful and accessible book.

Institute for Earth and Space Exploration,   Philip Stooke,
University of Western Ontario
London, ON, Canada
May, 2025

About the Author

Pamela Elizabeth Clark, PhD , grew up in New England and, inspired by President John Kennedy, decided to explore outer space by the time she was 10 years old. She thought, "If they can put a man on the moon, they can put a woman (me) on Mars!" She obtained her BA from St. Joseph College. There, she participated in organic geochemistry research with Dr. Mary Ellen Murphy, one of the developers of criteria for biogenicity for moon rocks, and coordinated an NSF inter-disciplinary undergraduate field research project. To obtain her PhD in Planetary Geochemistry (University of Maryland), she simulated, analyzed, correlated, and interpreted lunar X-ray spectra. She was a member of the group, led by Isidore Adler and Jack Trombka, that pioneered the use of orbital x-ray and gamma-ray spectrometers and participated in the USGS Astrogeology Branch Lunar Data Consortium, the first attempt to create a common format database for all the remote sensing data from a planetary body. After completing her PhD, she worked with the Goldstone Solar System Radar group, and expanded her remote sensing background to include radar, thermal and near infrared studies of planetary surfaces with particular emphasis on the study of Mercury's surface. Dr. Clark organized a briefing team to promote a mission to Mercury, for a while edited the *Mercury Messenger* newsletter, and published the (Springer) book *Dynamic Planet: Mercury in the Context of Its Environment*. Following this, she became a member of a GSFC group led by Steve Curtis that developed innovative concepts for autonomous, compact, low-cost multi-platform tools, instrument packages, and missions for extreme environments. At that point, Dr. Clark, along with Michael Rilee, published the (Springer) book *Remote Sensing Tools for Exploration*. Along with colleague Chuck Clark, she also published a book on Chuck's innovative mapping technique applied to planetary surfaces, *Constant-*

Scale Natural Boundary Mapping to Reveal Global and Cosmic Processes. More recently, her background and long-term interest in lunar exploration led her to support the development of science requirements, operational scenarios, and architectures for the return to the Moon as part of Project Constellation, which provided the basis for this book. She has also led the promotion of the CubeSat paradigm for deep space exploration in the planetary science community, developing the programs for the annual Lunar Cubes Workshops first sponsored by Flexure Engineering. She led the working group and was the Science PI for the NASA NextSTEP lunar orbiter Lunar IceCube built by the Space Science Center at Morehead State University and launched by ARTEMIS 1. Dr. Clark has worked at NASA GSFC and JPL, as well as at several universities during the course of her career and is currently the director of the Star Theater and member of the technical staff of the Space Science Center at Morehead State University.

Contents

1 **Overview: The Apollo Lunar Surface Expeditionary Approach**   1
   1.1 Context and Keys to Apollo's Success   1
   1.2 Representative Overviews   5
   References   28

2 **The Challenge: Constraints and Requirements of Extreme Exploration**   31
   2.1 Remote Exploration Before Apollo   31
   2.2 Lewis and Clark Expedition, 1803–1806 (Figs. 2.1, 2.2, 2.3, and 2.4)   32
   2.3 Stanley Trans-Africa Expedition, 1874–1877 (Figs. 2.5, 2.6, and 2.7)   37
   2.4 Scott, Shackleton, and Amundsen South Polar Expeditions, 1901–1916 (Figs. 2.8, 2.9, 2.10, 2.11, 2.12, and 2.13)   42
   2.5 Comparison to Apollo Lunar Expeditions   52
   References   57

3 **The Context: Robotic Precursors Ranger, Surveyor, Orbiter**   61
   3.1 Robotic Precursors   61
   3.2 The Ranger Program   62
   3.3 The Lunar Orbiter Program   69
   3.4 The Lunar Surveyor Program   77
   3.5 Pre-landing Crewed Apollo Missions   87
   References   96

4 The Preparation: Planning, Selecting, Training — 99
- 4.1 Planning and Simulating — 99
- 4.2 Astronaut Selection — 100
- 4.3 Astronaut Training Activities and Their Rationale — 108
- 4.4 1963–1968 Astronaut Training — 110
- 4.5 1969–1972 Apollo Prime and Backup Crew Field Training — 125
- 4.6 Field Simulations — 130
- References — 131

5 Field Work Approach — 133
- 5.1 Field Work Preparation — 133
- 5.2 Mobility on the Lunar Surface — 138
- 5.3 Field Work in a Space Suit — 141
- 5.4 Space Suit Design — 141
- 5.5 Field Documents and Documentation — 143
- 5.6 Documentation Devices: Maps and Cuff Checklists — 147
- 5.7 Field Work Practices — 152
- References — 153

6 Field Work Tools — 155
- 6.1 Sampling: The Primary Field Work — 155
- 6.2 The Toolkit — 158
- References — 168

7 Surface Deployed Experiments — 171
- 7.1 Apollo Lunar Science Experiment Packages (Fig. 7.1) — 171
- 7.2 Lunar Interior Experiments [8–13] — 176
- 7.3 Lunar Environment Experiments [8–13] — 187
- 7.4 Major Results from Experiments — 199
- References — 200

8 Apollo Surface Missions — 205
- 8.1 Context — 205
- 8.2 Early Missions: Apollo 11 and 12 — 205
- 8.3 Transitional Mission Apollo 14 — 209
- 8.4 The Apollo J Missions — 210
- 8.5 Apollo 15 Mission (Figs. 8.4 and 8.5) — 211

8.6	Apollo 16 Mission (Figs. 8.6 and 8.7)	214
8.7	Apollo 17 Mission (Figs. 8.8 and 8.9)	218
8.8	Understanding and Integrating Apollo J Mission Results	221
8.9	The Cancelled Missions: Apollo 18, 19, and 20	222
8.10	Next Generation Missions	226
References		242

9 Beyond Apollo: Applying Lessons Learned — 247
- 9.1 Once and Future Lunar Exploration: Future — 247
- 9.2 In Situ Exploration Tools and Compact Package Networks — 250
- 9.3 State-of-the-Art Documentation — 252
- 9.4 Power and Thermal Challenges — 252
- 9.5 Next Generation Rovers — 255
- 9.6 Advanced Autonomy for Locomotion — 261
- 9.7 Advanced Autonomy for Dynamic Structures — 264
- 9.8 Communication and Navigation Challenges — 266
- 9.9 Dust Mitigation Challenges — 269
- 9.10 State of the Art Astronaut Gear — 273
- 9.11 Transportation Services: Launch, Lunar Orbit and Landing — 274
- 9.12 In Situ Manufacturing and Resource Extraction — 275
- References — 276

Afterword: Final Thoughts — 283

Index — 287

List of Figures

Fig. 1.1 Iconic view of Apollo 15 astronaut Commander David Scott saluting the American flag deployed at the Apollo 15 landing site clearly showing Hadley Delta in the background. (Image Credit: NASA JSC)   2

Fig. 1.2 EVA at the Apollo 12 landing site. Apollo 12 astronaut Alan Bean filling sample tube. Astronaut Pete Conrad can be seen reflected in his visor. (Image Credit: NASA/MSFC History Office)   4

Fig. 1.3 Farouk El Baz discussing landing sites. (Image Credit: NASA HQ History Office)   5

Fig. 1.4 Farouk El Baz reviewing maps with Ronald Evans and Robert Overmyer. (Image Credit: NASA HQ History Office)   6

Fig. 1.5 Lee Silver points out salient feature to Charles Duke and John Young during field training. (Image Credit: NASA and USGS)   9

Fig. 1.6 Jim Head at NASA 50th Anniversary celebration at Langley Research Center in 2008. (Image Credit: NASA LRC)   14

Fig. 1.7 Harrison Schmitt training in the Lunar Module simulator. (Image Credit: NASA KSC)   21

Fig. 2.1 Lewis and Clark Expedition Paraphernalia. Top, Clockwise from upper left: sextant, compass, pocket telescope [5]. Bottom: Ink and writing tools [6]. (Image Credits: National Park Service Photos/Kephart)   32

Fig. 2.2 Detail of Paxson (1912) Lewis, Clark, and Sacajawea at Three Forks, Lobby of Montana House of Representatives [7]. (Source: Public Domain)   33

Fig. 2.3 Lewis and Clark Expedition. Top: Blunderbuss from that era [8]. Image Credit: Smithsonian, National Museum of American History. Bottom: Map of Lewis and Clarks Track along the

	Missouri, across the Rocky Mountains, down the Columbia River, to the Pacific Ocean. Clark, History of the Expedition under Command of Captains Lewis and Clark, to the Sources ot he Missouri thence across the Rocky Mountains and Down the River Columbia to the Pacific Ocean 1814 [9]. (Image Credit: Library of Congress)	34
Fig. 2.4	Lewis and Clark Expedition. The Great Falls, 'the grandest site sight he had ever seen according to Lewis [10]. (Credit: Montana Historical Society. Public Domain)	35
Fig. 2.5	Explorer Henry Morton Stanley with Kalulu, his gun bearer and servant on expeditions and otherwise constant companion who became his adopted child. Stanley is posing in suit he wore when he met Livingstone [14]. (Source: Public Domain)	38
Fig. 2.6	Stanley's routes indicated by heavy black line in Circumnavigation of Lake Victoria found to be the source of the Nile via the White Nile tributary and Lake Albert in the upper left corner (top map) and Exploration of Lake Tanganyika (bottom map). The Lady Alice was used to transport them by water [14]. (Source: Public Domain)	39
Fig. 2.7	Map of Central African area of Stanley's expedition showing relationship to Nile and Congo Rivers with the divide between Congo and Nile Basins in green [15]. (Source: {annotated} Public Domain)	40
Fig. 2.8	Top: Shackleton expedition with dog sled on the ice [30]. Source public doman. Bottom: Shackleton expedition with Mongolian ponies on the ice [31]. (Source: Public Domain)	42
Fig. 2.9	Top: Shackleton's Base Camp where officers and crew shared common space having an evening's entertainment [32]. Source public domain. Bottom: Some of Shackleton's men on the ice sheet [33]. (Source: Public Domain)	43
Fig. 2.10	Top: Scott's Crew at an Ice Depot [34]. Source public domain. Bottom: Scott in Officers' quarters were kept separate from the crews' quarters [35]. (Source: Public Domain)	44
Fig. 2.11	Top: Decorated Interior of Amundsen Base Camp [36]. Source public domain. Bottom: Amundsen Base Camp [37]. (Source: Public Domain)	45
Fig. 2.12	Top: Olaf Blaaland, one of Amundsen's crew, in their arctic gear made of natural and-time-tested material [38]. Bottom: Amundsen camp at South Pole [39]. (Source: Public Domain)	46

Fig. 2.13	Routes of Amundsen, Shackleton, and Scott [40]. (Source: Public Domain)	47
Fig. 2.14	Distinctive shape, size distribution, and components of lunar soil as discussed in the text [56]. (Credit: NASA Lunar Initiative Briefings)	55
Fig. 3.1	Ranger Two at JPL. (Credit: NASA/JPL-Caltech)	62
Fig. 3.2	Ranger 7 Mission Details. (Credit: NASA/JPL-Caltech)	63
Fig. 3.3	Top: Ranger 9 Approaching the Moon (Top left to right, then bottom left to right. (Credit: NASA/JPL-Caltech). Bottom: First successful Ranger lander (7) landing site. (Courtesy of NASA)	64
Fig. 3.4	JPL Mission Control for Ranger Program 1964. (Credit: NASA/JPL-Caltech)	65
Fig. 3.5	Ranger Block 3 (7, 8, and 9) TV cameras. Wide Angle cameras located in center in both banks. (Credit: NASA/JPL-Caltech)	67
Fig. 3.6	Top: Lunar Orbiter Spacecraft. Bottom: Lunar Orbiter Camera System [10]. (Credits: NASA/NSSDC)	69
Fig. 3.7	Labelled Schematic of Lunar Orbiter Spacecraft in two orientations. (Credit: NASA)	70
Fig. 3.8	Top: Iconic oblique view of Crater Copernicus by Lunar Orbiter 2. Bottom: Lunar Reconnaissance Orbiter Map View of Copernicus for Comparison. (Credit: NASA)	71
Fig. 3.9	Iconic first view of Earth from the Moon by Lunar Orbiter 1, August 23, 1966. Top is original image. Bottom is processed by Lunar Orbiter Imager Recovery Project (LOIRP) [12]. (Credit: NASA)	72
Fig. 3.10	Panoramic mosaics of Survey 6 (top) and Survey 7 (bottom) landing sites. (Credit: NASA)	78
Fig. 3.11	Surveyor 6 Footpad 2 (NASA) showing rocky portion of surrounding regolith. (Credit: NASA)	79
Fig. 3.12	Surveyor 1 shadow. (Credit: NASA)	80
Fig. 3.13	Surveyor Lander Schematic. (Credit: NASA)	81
Fig. 3.14	Surveyor 3 picture taken by Apollo 12 astronaut (top) and with Apollo 12 astronaut nearby and Landing Module in background. (Credit: NASA)	82
Fig. 3.15	Surveyor Scooping arm (top) and Alpha backscatter instrument (bottom) deployment mechanisms. (Credit: NASA)	83
Fig. 3.16	Top: Apollo 7 Crew in first live Television broadcast from space. Bottom: Mission Control Houston during Apollo 10 mission. (Credits: NASA)	88
Fig. 3.17	From Top to Bottom Apollo 10 astronauts Schirra, Eisele, and Cunningham in orbit. (Credit: NASA)	89

Fig. 3.18 From Top: Schirra, Eisele, Cunningham after Apollo 7 splashdown on deck of Essex. Bottom: Kranz, Lunney, and Griffin celebrating in mission control after Apollo 7 splashdown. (Credits: NASA) — 90

Fig. 3.19 From Top: Marilyn Lovell (center) accompanied by Alan Shepard (left) in Mission Control Visitors' Gallery. Bottom: Apollo 8 capsule photographed at 40,000 feet during reentry by support crew. (Credits: NASA) — 91

Fig. 3.20 Iconic Earthrise over Moon taken by Bill Anders during lunar obit of Apollo 8. (Credits: NASA) — 92

Fig. 3.21 Top: Astronaut Schweikart on Lunar Module Spider's front porch evaluating PLSS backpack during Apollo 9 Earth orbit. Bottom: Astronaut Scott supporting the undocking of Spider from Service Module Gumdrop during Apollo 9 Earth orbit. (Credits: NASA) — 93

Fig. 3.22 Apollo 9 Lunar Module flown by astronaut Schweickart in Earth Orbit. (Credits: NASA) — 94

Fig. 3.23 Top: Apollo 10 on crawler between VAB and launchpad. Bottom: Apollo 10 Lunar Module Snoopy being tested in Lunar Orbit by Stafford and Cernan. (Credits: NASA) — 95

Fig. 3.24 Support Crew inspects egress slide wire gondola before Apollo 10 mission. (Credit: NASA) — 96

Fig. 4.1 Project Mercury astronauts. top: Cooper, Carpenter, Glenn, Shepherd, Grissom, Schirra, Slayton in training; bottom: Grissom, Shepard, Carpenter, Schirra, Slayton, Glenn, Cooper with rocket. (Courtesy of NASA) — 103

Fig. 4.2 Jerrie Cobb, one of the 'Mercury 13' as described in the text who passed astronaut screening successfully and then were denied access to the astronaut program. Top: Jerrie standing by Mercury capsule. Bottom: Jerrie in NASA Multi-Axis Space Test Inertia Facility (MASTIF). (Courtesy of NASA) — 104

Fig. 4.3 Second and third group of astronauts: Top: New Nine selected for Gemini: L to R Back See, McDivitt, Lovell, White Stafford. L to R Front: Conrad, Borman, Armstrong, Young. Bottom: The Fourteen for Gemini and Apollo: L to R Back: Collins, Cunningham, Eisele, Freeman, Gordon, Scheickart, Scott, Williams. L to R Front Aldrin, Anders, Bassett, Bean, Cernan, Chafee. (Courtesy of NASA) — 106

Fig. 4.4 Top: The Scientists (Group 4): L to R Front: Michel (physicist), Schmitt (astrogeologist), Kerwin (physician). L to R back: Garriott (Engineer), Gibson (Engineer). Graveline (physician) missing. Bottom: Original Nineteen

	(Group 5): L to R back: Swigert, Pogue, Evans, Weitz, Irwin, Carr, Roosa, Worden, Mattingly, Lousma. L to R Front Givens, Mitchell, Duke, Lind, Haise, Engle, Brand, Bull, McCandless. (Courtesy of NASA)	107
Fig. 4.5	Top: Excess 11 (Group 6): L to R front Chapman, Parker, Thornton, Llewellyn. L to R back Allen, Heinze, England, Holmquest, Musgrave, Lenoir, O'leary. Bottom: Quarantine box for Apollo 11 rocks being unloaded at Ellington Air Force Base near Receiving Lab. NASA 'Official Witnesses' George Low, Samuel Phillips, Thomas Paine, and Robert Gilruth pose with the box. (Courtesy of NASA)	109
Fig. 4.6	USGS 'Explorer' vehicle used to transport astronauts around cratered sites. (Courtesy of USGS)	118
Fig. 4.7	Cinder Lake Crater Fields created to simulate the lunar surface. Note simulated LM location in Crater Field 1. (Courtesy of Phil Stooke, Personal Communication)	118
Fig. 4.8	Cernan and Schmitt at Nevada test site. (Courtesy of NASA)	119
Fig. 4.9	Unidentified trainer and astronaut at Grand Canyon. (Courtesy of NASA and USGS)	120
Fig. 4.10	Cernan at Big Bend. (Courtesy of NASA)	121
Fig. 4.11	Unidentified astronauts at Newberry Crater, Oregon with local expert Aaron Waters (holding box) discussing origins of volcanic features. (Courtesy of NASA)	121
Fig. 4.12	Fred Haise and Jim Lovell at Big Island pahoehoe lava flow. (Courtesy of NASA)	123
Fig. 4.13	Two astronauts at Katmai Valley. (Courtesy of NASA)	123
Fig. 4.14	Astronauts Vance Brand and Stuart Roosa (left) with Iceland volcanics instructor (glasses) Ted Ross in Askja, Iceland. (Courtesy of NASA)	124
Fig. 4.15	Astronauts Ed Mitchell and Stuart Roosa learning to use a gravity meter at Pinacates volcanic region, Sonora, Mexico. (Courtesy of NASA)	124
Fig. 4.16	Haise and Lovell (top), Armstrong and Aldrin (bottom) at Sierra Blanco. (Courtesy of NASA)	125
Fig. 4.17	At Mono Craters site, Instructor Muelhberger (left) talks with Astronaut Young in the foreground. Behind them, Instructor Wones talks with Astronaut Duke. (Courtesy of NASA)	126
Fig. 4.18	Astronauts Duke and Young at Sudbury Basin. (Courtesy of NASA)	127

List of Figures

Fig. 4.19 Unidentified Astronauts at Crater of the Moon, Idaho. (Courtesy of NASA) — 128

Fig. 4.20 Astronaut Schmitt at the Rio Grande Gorge near Taos, NM. (Courtesy of NASA) — 128

Fig. 4.21 Black Hawk Landslide in foreground. (Courtesy of NASA/METI/AIST/Japan Space Systems and US/Japan ASTER Science. Team) — 129

Fig. 4.22 Gene Cernan and Harrison Schmitt on EVA practice in Nevada Volcanic Field near Tonopah. (Courtesy of NASA) — 130

Fig. 5.1 Gene Shoemaker dons rocket belt during field training. (Courtesy USGS) — 139

Fig. 5.2 The main components of the Lunar Roving Vehicle. (Courtesy of NASA) — 140

Fig. 5.3 (**A**) Apollo spacesuit with EVA features [11]. (**B**) Apollo 16 astronaut Duke's glove [11]. Note Thermal Micrometeorite Shield and reinforced fingertips [14]. (**C**) Glove components: Astronaut puts on Comfort glove, then IV pressure glove with bladder and inner restraint, then EV outer glove with convolute wristlet (to allow movement) and shield (seen in B) [12, 13]. (Courtesy of NASA) — 142

Fig. 5.4 Apollo 15 Surface Journal Excerpt [17]. (Courtesy of NASA) — 145

Fig. 5.5 Apollo 16 Surface Journal Excerpt [18]. (Courtesy of NASA) — 146

Fig. 5.6 Apollo 17 Surface Journal Excerpt [19]. (Courtesy of NASA) — 147

Fig. 5.7 Schedule and cuff check list for Apollo 16 Station 4 Stone Mountain. Note nature of schedule and mnemonic nature of checklist as described in text [18]. (Courtesy of NASA) — 148

Fig. 5.8 Apollo 17 astronaut Cernan with cuff check list, closeup on the right. (Courtesy of NASA) — 149

Fig. 5.9 Apollo 14 Astronaut Mitchell using map. (Courtesy of NASA) — 149

Fig. 5.10 Apollo 15 Lunar Navigation map used on lunar surface. (Courtesy of NASA) — 150

Fig. 5.11 Apollo 16 Lunar Navigation map used on lunar surface. (Courtesy of NASA) — 151

Fig. 5.12 Apollo 17 Lunar Navigation map used on lunar surface. (Courtesy of NASA) — 152

Fig. 6.1 Large tool carrier used on Apollo J missions with some tools displayed below left to right: core tube caps, documented sample bags, hammer, drive tube caps, two pairs of tongs, adjustable handle scoop and extension handle [11]. (Courtesy of NASA JSC) — 160

Fig. 6.2 Sample Collection Tools [11] clockwise from top: Small adjustable-angle scoop attached to long extension

	handle drawing (left) and being used during Apollo 16 (right). Bottom: Rake sample collecting rock fragments during Apollo 16. (Courtesy of NASA JSC)	161
Fig. 6.3	Sample collection tools [11] from top to bottom: Trenching tool with adjustable angle blade, Apollo J mission heavier hammer, 32-in. tongs [11]. (Courtesy of NASA JSC)	162
Fig. 6.4	Sample documentation equipment [11]. Top left 20-bag dispenser for flat, rectangular labeled bags on Apollo J missions. Right one of the bags opened in the lab to show Apollo soil sample 74,220. The aluminum rims hods bag open. Bottom: three 20 bag dispensers packed inside sample collection bag prior to flight. (Courtesy of NASA JSC)	163
Fig. 6.5	Sampling equipment [11]. Top: Cup-shaped labeled sample bags used in LRV soild smaple on Apollo 17. Middle: Apollo 14 sample return container prior to flight with round documented sample bags, 2 cm diameter core tubes, core tube caps, and magnetic shield sample container (white cylinder). Bottom left Gnomon at Apollo 17 site, right sample scale. (Courtesy of NASA JSC)	164
Fig. 6.6	Surveying Staff being field tested with Lunar Geological Exploration Camera (LGEC) (top left and right), with laser range finding reflector and spectrographic systems as described in the text [12]. (Courtesy of USGS)	165
Fig. 6.7	Apollo Geology Tools. Top Proposed Desktop-Sized thin section and slicing machine from Microtek that never flew due to limitations in allowable equipment mass. Bottom: 2-cm diameter core tube attached to shorter style extension handle being driven into the regolith on Apollo 12. (Courtesy of NASA)	166
Fig. 6.8	Power Drill used on the Apollo J missions being set up by Astronaut Irwin on Apollo 15 (top) with drawing of major components below [11]. (Courtesy of NASA JSC)	167
Fig. 7.1	Top: Labeled drawings of many ALSEP instruments [2]. Bottom: ALSEP package deployed on Apollo 16 with Magnetometer in foreground and Astronaut Young approaching the central station in the background [2]. (Courtesy of NASA)	172
Fig. 7.2	Top: Schematic of the ALSEP Central Station on the left which distributed power from the RTG on the right [2]. Bottom: Apollo 12 astronaut Bean offloading the ALSEP [2]. (Courtesy of NASA)	175
Fig. 7.3	Laser retroreflector deployed (above) and drawn with label parts (below) [2]. (Courtesy of NASA)	177

Fig. 7.4	Seismic Experiments. Passsive Seismometer (top) and active seismometer components with explosive packages (closeup on right) mounted on pallet to be deployed along a line of geophones [2]. (Courtesy of NASA)	178
Fig. 7.5	Distribution of 4 geophones round the LSPE lunar seismic profiling experiment on Apollo 17 [2]. (Courtesy of NASA)	179
Fig. 7.6	Heat Flow Experiment Schematic of proper deployment finally accomplished on Apollo 17 (top). Below, first probe successfully deployed in hole made by power drill on Apollo 17 [2]. (Courtesy of NASA)	182
Fig. 7.7	Drawing of the portable Traverse Gravimeter Experiment (TGE) (above) deployed on the lunar surface (below) [2]. (Courtesy of NASA)	184
Fig. 7.8	Traverse Gravimeter Experiment (blue gray box) mounted on the back of the Apollo 17 rover. Astronaut Cernan deployed and set up the TGE, took measurements, and remounted it at each station. He took this shot to document the 'replacement fender' made from the back of the manual [2]. (Courtesy of NASA)	185
Fig. 7.9	Top: Schematic of the Magnetometer. The magnetometer deployed on the surface can be seen in Fig. 7.1 [2]. (Courtesy of NASA). Bottom: Lunar Neutron Probe schematic cross-section illustrating activated (boron targets and mica detectors on central rod face plastic detectors and uranium targets on rib cage) mode on left and deactivate (targets and detectors do not face each other on right [13, 34]. (Courtesy of NASA Apollo 17 preliminary science report)	186
Fig. 7.10	Dust Detectors. Lunar Dust Detector (LDD) deployed on the lunar surface (top left) and Lunar Ejecta And Meteorites (LEAM) in the lab (top right) [2]. Courtesy of NASA. Bottom: Schematic of the LDD dust detector. (Courtesy of NASA/NSSDC)	188
Fig. 7.11	Solar Electrical Properties (SEP) Experiment being deployed on the lunar surface by Apollo 17 astronaut Schmitt (top) and a closeup with labels before deployed [2]. (Courtesy of NASA)	190
Fig. 7.12	Cosmic Ray Detector (CRD) (left) and Solar Wind Composition Experiment (SWC) (right) as described in the text [2]. (Courtesy of NASA)	191
Fig. 7.13	Solar Wind Spectrometer (SWS) schematic (top) and after deployment on Apollo 15 (bottom) [2]. (Courtesy of NASA)	192

Fig. 7.14	Charged Particle Lunar Environment Experiment (CPLEE). Interior schematic of charged particle analyzer and sketch showing fields of view and look direction (top) and deployment on Apollo 14 [2]. (Courtesy of NASA)	194
Fig. 7.15	Lunar Atmospheric Composition Experiment (LACE). Deployment on Apollo 17 (top) and schematic of magnetic deflection mass spectrometer (bottom) [2]. (Courtesy of NASA)	195
Fig. 7.16	Cold Cathode Ion Gauge (CCIG) and Suprathermal Ion and Mass Analyzer Detector SIDE for Apollo 15 mission reflecting significant redesign from earlier missions as described in text. CCIG required input from SIDE to operate [2]. (Courtesy of NASA)	197
Fig. 8.1	Apollo 11 Landing Site Map. LM and Flag indicate the location of the lander and the flag respectively. The five panoramic camera views (Pan), Hasselblad camera images (Apollo Atlas Image numbers beginning with AS11), the TV camera (TV), and EASEP instrument (SWC, LRRR, PSE, CDR) deployment sites indicated. Two major landmarks, Double and Little West Crater, are also indicated. (Courtesy of USGS Astrogeology Branch)	207
Fig. 8.2	Apollo 12 Landing Site Map. Locations of the lander (LM), ALSEP, and Surveyor 3 are indicated, along with major physical landmarks labelled. The EVA1 and EVA2 traverses follow the red lines. Panoramic camera views (Pan), Hasselblad camera images (Apollo Atlas Image numbers beginning with AS12), HD (high definition), MD (medium definition), photos, and local photos (SP, HD, BK, BN) for Sharp Crater, Halo Crater, Block Crater, and Bench Crater) indicated. (Courtesy of USGS Astrogeology Branch)	208
Fig. 8.3	Apollo 14 Landing Site Map. Locations of the lander (LM), ALSEP with Geophone line, LRRR, Bulk, Comprehensive, Grab, and Contingency samples are indicated, along with major physical landmarks labelled. The EVA1 and EVA2 traverses follow the red lines. A, B, C, D, E, F, G, and H indicate documentation taken sequentially along the second traverse. (Courtesy of USGS Astrogeology Branch)	209
Fig. 8.4	Apollo 15 traverses (left), orbital image (top right) and ground photo(middle right) of landing site. Most significant samples include (left to right along bottom) Imbrium basin impact melt rock, the oldest piece of crust collected called 'genesis rock', and young very low titanium flood basalt, as well as (insert above), the	

	first pyroclastic material discovered, green glass [11]. (Courtesy Clark, Fig. 62,010, Earth Moon and Planets)	212
Fig. 8.5	Schematic Geological Cross–Section of the Apollo 15 Landing Site Area from SSE to NNW (modified from [22]) illustrating the complex nature of the interface between underlying crustal structures generated by impact represented by Hadley Delta, crustal impact debris generated by several major impacts, and episodic volcanic plains formation subsequent to Imbrium basin formation [11]. (Courtesy of Clark, 2010, Earth Moon and Planets)	213
Fig. 8.6	Apollo 16 traverses (left), orbital image (top right) and ground photo (middle right) of landing site. Typical rocks, include breccia (left) with clasts containing dark volcanic material, impact melt (middle), and bright crustal fragments (extreme right) [11]. (Courtesy of Clark, Fig. 8, 2010, earth moon and planets)	215
Fig. 8.7	Schematic Geological Cross–Section of the Apollo 16 Landing Site Area from N to S (modified from [22]) illustrating the complexity introduced by repeated excavation and deposit of impact ejecta and melt of highland crust, as well as the stratigraphic relationship between the younger Cayley Formation, the surrounding Descartes Formation, and an even older underlying formation [11]. (Courtesy of Clark, Fig. 9, 2010, earth moon and Planets)	216
Fig. 8.8	Apollo 17 traverses (left), oblique orbital image (top right) and ground photo (middle right) of landing site. Typical rocks include (bottom left to right) crustal (highland rocks) like olivine–bearing troctolite (left), norite (middle), and the pyroclastic find (orange glass) [11]. (Courtesy of Clark, Fig. 10, 2010, earth moon and planets)	218
Fig. 8.9	Schematic Geological Cross–Section of the Apollo 17 Landing Site Area from SW to NE (modified from [22]) illustrating the complexity of the interface between mare filled graben Littrow Valley and surrounding mountains of Serenitatis rim, as well as the distribution of the diverse materials found in the area, including light mantle (South Massif landslide) and dark mantle overlying older mare basalt in the valley, crustal ejecta and breccias of the massifs, and the Sculptured Hills material [11]. (Courtesy of Clark, Fig. 11, 2010, earth moon and planets)	219
Fig. 8.10	Apollo 18 Proposed Landing Site Photos. Top: Schroter's Valley as described I the text, with Cobra's Head (and its vent) at the left end. Bottom: Proposed Landing site (circled)	

	at lower (left) and higher (right) resolution. Note Landing site is adjacent to large vent on Cobra's Head. (Photos courtesy of NASA)	225
Fig. 8.11	Proposed Apollo 19 Proposed Landing Site Photos. Note Marius Hills extensive and varied volcanic feature and the presence of a magnetic swirl anomaly as described in text. The insert in the bottom right is a close up of the proposed landing site surrounded by four domes plotted in red on the larger map. (Photos courtesy of NASA)	226
Fig. 8.12	Proposed Apollo 20 Landing Site Photos. Top left is recent LRO LROC view of the Copernicus Crater area as described in the text. The insert to the right is a closeup of the central peaks where boulders downslope of the central peak can clearly be seen. Bottom is the closeup of the proposed landing site just north of the crater's central peak. (Photos courtesy of NASA)	227
Fig. 8.13	Lunar Orbiter Photo of Marius Hills showing diversity of features including volcanic cones and domes, mare Flows, sinuous rilles, straight rilles (faults?), wrinkle ridges, impact craters, secondary crater chains, and magnetic swirl anomaly Reiner Gamma (bottom center). (Photo Courtesy of NASA)	232
Fig. 8.14	Geological map derived from photo seen in Fig. 13 with rilles, ridges, crater chains, and Reiner Gamma marked. 10 km radius circles indicate two potential landing sites (Marius West and Marius East) identified in study described in text [46]. (Courtesy of Clark et al., Fig. 5, 2010, GSA Bulletin)	232
Fig. 8.15	Marius East landing site with traverses, as described in text, marked in red. Visits planned to all identified morphologic and spectral units, enables analysis of unit contacts and organization of local developmental history including insight into impact, tectonic, volcanic, regolith, and magnetic processes [46]. (Courtesy of Clark et al., Fig. 8, 2010, GSA Bulletin)	233
Fig. 8.16	Marius West landing site with traverses, as described in text, marked in red. Visits planned to all identified morphologic and spectral units, enables analysis of unit contacts and organization of local sequential history, with visits to both sides of a volcanic rille and magnetic swirl anomaly, providing insight into impact, volcanic, and regolith processes [46]. (Courtesy of Clark et al., Fig. 9, 2010, GSA Bulletin)	233
Fig. 8.17	Landing Site Selection as a function of study area size [46] (Fig. 6). Sites were selected on the basis of providing the	

	best access to the widest range of volcanic and tectonic features representing the formation and history of Marius Hills in the context of the western nearside. Within study areas of given radii 10, 20, and 30 km, respectively. Two 10 km radius (20 km dimeter) and two 20 km radius sites met most of those criteria. Each 30 km radius site met those criteria, which included the magnetic swirl anomaly, but with different emphases. This study focused on the 10 km radius sites because they best fit the 'sortie' architecture under consideration [46]. (Courtesy of Clark et al., 2010, Fig. 6, GSA Bulletin)	234
Fig. 8.18	Tsiolkovsky Basin of the lunar farside as seen from Lunar Orbiter. (Photo courtesy of NASA)	236
Fig. 8.19	Tsiolkovsky Basin regional scale scenario with selected peak structural study features for central peak and interior routes as described in the text [47]. (Courtesy of Clark et al., Fig. 8a, 2010, GSA Bulletin)	237
Fig. 8.20	Tsiolkovsky Basin regional scale scenario with selected mare basin and rim study features as described in the text [47]. (Courtesy of Clark et al., Fig. 8b, 2010, GSA Bulletin)	238
Fig. 8.21	Geological unit map of the Lunar Farside South Pole Aitken Basin with potential traverse routes superimposed. Note variations in major terrains, with less cratered terrain within the basin and dark plains implying cryptomare distributed largely in the northern portion. Landmarks discussed in text are marked with stars [47]. (Courtesy of Clark et al., Fig. 4a, 2010, GSA Bulletin)	239
Fig. 8.22	Regional scale mission traverse routes from Apollo Basin outpost, with close-ups of feature of primary interest, Apollo Basin, Olivine Hill, and Oppenheimer pyroclastics [47]. (Courtesy Clark et al., Fig. 7a, GSA Bulletin)	240
Fig. 8.23	Regional scale mission traverse routes from Thomson Crater outpost, with close-up of feature of primary interest, Mare Ingenii magnetic swirl anomaly [47]. (Courtesy of Clark et al., Fig. 7b, 2010, GSA Bulletin)	241
Fig. 8.24	Regional scale mission traverse routes from Shackleton Crater outpost, with close-up of feature of interest Schrodinger Basin [47]. (Courtesy of Clark et al., Fig. 7c, 2010, GSA Bulletin)	241
Fig. 9.1	JPL CADRE mini-rover with Mars Perseverance rover engineering model in background. (Credit: NASA JPL-CalTech)	256
Fig. 9.2	NASA Concepts for unpressurized rovers from left to right Lunar Outpost Eagle, Moon RACER, Venturi Astrolab FLEX. (Credit: NASA/Bill Stafford)	257

Fig. 9.3	NASA Concepts for pressurized rover: JAXA Cruiser. (Credit: NASA and JAXA)	257
Fig. 9.4	Intuitive Machines proposed Micro-lander 'Nova-C' capable of multiple hops on the lunar surface as described in the text [29]. (Credit: Intuitive Machines)	258
Fig. 9.5	Model of ETH Zurich battery powered gimballed-leg Tumbler capable of multiple jumps as descried in the text [31]. (Credit: Evans, IOT World Today)	259
Fig. 9.6	MIT proposed Hovercraft with Ion Propulsion Drive for exploring the lunar surface as described in the text [30]. (Credit: Catwell, Element 14)	259
Fig. 9.7	Model of Festo hopping kangarobot as described in the text [32]. (Credit: Del Guercio, Science Today)	260
Fig. 9.8	NASA/GSFC Concept for 3D reconfigurable tetrahedral rover as described in text. (Credit: NASA/AMA Productions)	261
Fig. 9.9	Concept of reconfigurability and multifunctionality of ANTS-based robotics. Top: Animation of EMS (macro) level tetrahedral rover (left to right) climbing, bridging, flattening, and clambering over obstacles in natural terrain. Bottom: Animation of MEMS level ANTS-based dynamic infrastructure (left to right) landing, maneuvering around obstacles, communication. (Credit: NASA/RASC project)	265
Fig. 9.10	NASA/GSFC Concept of SPARCLED handheld electrostatics-based dust removal tool as described in text [54]. (Modified from Clark et al., Fig. 8, 2010, AIP Conference Proceedings)	272

List of Tables

Table 2.1	Apollo program and its precursors	53
Table 3.1	Apollo program and its precursors	61
Table 3.2	Ranger missions	66
Table 3.3	Lunar orbiter missions	75
Table 3.4	Successful lunar surveyor landers	84
Table 4.1	Astronauts selected 1959–1969	102
Table 4.2	Astronaut field training 1963 and 1964	111
Table 4.3	Astronaut field training 1965	112
Table 4.4	Astronaut field training 1966, 1967, and 1968	113
Table 4.5	Astronaut field training, assigned Apollo 11–14 prime and backup crews 1969	114
Table 4.6	Astronaut field training, assigned Apollo 13–16 prime and backup crews 1970	114
Table 4.7	Astronaut field training, assigned Apollo 15–17 prime and backup crews, 1971	116
Table 4.8	Astronaut field training, assigned Apollo 16 and 17 prime and backup crews, 1972	117
Table 5.1	Apollo Documentary Film Table 1	134
Table 5.2	Apollo Documentary Film Table 2	134
Table 5.3	Apollo Documentary Film Table 3	135
Table 5.4	Apollo Documentary Film Table 4	135
Table 5.5	Apollo Documentary Film Table 5	136
Table 5.6	Apollo Documentary Film Table 6	136
Table 5.7	Apollo Documentary Film Table 7	137
Table 5.8	Apollo Documentary Film Table 8	138
Table 5.9	Systematic oral documentation of lunar surface geology	144
Table 6.1	Time and resource commitments for lunar surface activities	157

Table 6.2	Apollo Geological Toolkit	159
Table 7.1	Apollo surface deployed experiments [2]	173
Table 7.2	ALSEP 'Observatory' measurement objectives [3]	173
Table 8.1	Typical 'top ten' future lunar landing sites	228
Table 8.2	Apollo-style sorties equipment done currently for 2 groups of 2 astronauts	229
Table 8.3	Sortie-level operational scenarios	231
Table 8.4	Regional scale studies equipment beyond sorties in Table 8.2	235
Table 8.5	Regional scale operational scenarios	235
Table 9.1	Post-Apollo era successful American lunar missions or payloads	249
Table 9.2	Uncrewed lunar rovers	255
Table 9.3	JPL robotic rovers	256
Table 9.4	NASA Artemis rover concepts	258
Table 9.5	Essential regolith characteristics	269

1

Overview: The Apollo Lunar Surface Expeditionary Approach

1.1 Context and Keys to Apollo's Success

The extraordinary challenges faced in planning the first human expeditions to the surface of another solar system body led to the development of a new and distinctive approach to exploration in general and scientific field work in particular [1–8]. Those involved had to deal effectively with the extreme limitation in access to and resources available for a target as remote as the lunar surface, while developing a truly rigorous and effective approach to science activities ranging from collecting samples to deploying field instruments. As the results ranging from the truly representative lunar sample collection to the first measurements of the lunar environment indicate, the ground-breaking approach developed in response to these challenges was successful (Fig. 1.1).

The principal characteristics and keys to success of this approach, how these affected the planetary exploration that followed, and implications for future exploration, will be discussed in this book. These include

- highly integrated, intensive, and lengthy science planning, simulation, and astronaut training;
- development of a systematic scheme for selection and characterization of geological sites;
- a flexible yet disciplined methodology for sample documentation and collection.

In addition, the capability for constant communication with a 'backroom' of geological experts who make requests and weigh in on surface operations was

Fig. 1.1 Iconic view of Apollo 15 astronaut Commander David Scott saluting the American flag deployed at the Apollo 15 landing site clearly showing Hadley Delta in the background. (Image Credit: NASA JSC)

innovative and very useful in encouraging rapid dissemination of information to the greater community in general.

Online and hardcopy archives of the Apollo era science activity related documents (see references below) provide evidence of the nature of the approach, including the following collections in particular. The most detailed hardware specifications, which used to be available through 'one stop shopping' at the JSC websites, are no longer being maintained there, but can still be found a bit more dispersed at other locations.

- JSC History Collection at the University of Houston Clearlake [9]. Consists of portions of Apollo materials scanned into pdf files by JSC and now available as electronic files via reference request. Request for access can be made by contacting the UHCL archivist or the JSC History Office. Description of surface experiments and links to Apollo Surface Journal are available.
- Encyclopedia Astronautica descriptions of activities and equipment under the heading of 'Apollo' [10].
- Lunar Sourcebook [11] Most comprehensive analysis of science and engineering results from observations taken during all Apollo missions
- National Archives and Records Administration, Fort Worth Regional Archives. In theory, the Apollo portion of JSC History Collection is available in hard copy at this facility. However, my experience has been that the

Fort Worth Regional Office is so concerned about ITAR restrictions (on >50 year old obsolete technology!) that even an American citizen will not get access to this material from them. I recommend first attempting to obtain permission by contacting the JSC History Office rather than going directly to the Regional Office. On the other hand, an online search of 'NASA Apollo Archives' at the NARA website did turn up over a thousand references to small collections at various Archive offices and Presidential Libraries [12].

- Smithsonian Air and Space Museum has online Apollo Program pages including summaries of Apollo missions with historic data on each crew, briefs on Apollo vehicles and launches, and an Apollo surface tool collection [13].
- NSSDC (GSFC) includes descriptions of surface activities and experiments for each Apollo mission [14].
- NASA Headquarters has a catalogued list of Apollo hard copy documents (that can be seen by appointment) as well as Apollo program documents for each mission [15, 16].
- No study of the Apollo program activities would be complete with access to the Apollo Surface Journal [17] and Apollo Flight Journal [18], transcriptions of audio recording of the Apollo astronauts while they were in lunar orbit or on the lunar surface. The Apollo Surface Journal references samples collected and pictures taken by catalogue number.
- Extensive descriptions of all tools used on the lunar surface are available in the online geological tool catalog [19]. Details of all experiments deployed on the lunar surface are also available online [20].
- Although the original site for the details of lunar rover specification and performance documentation are no longer available at the JSC history collection website, much of this material can be found at sites [21–23].
- An extensive collection of oral history interviews of a broad cross-section of participants in the Apollo program, some of them presented and discussed here, may be found in the JSC oral history collection [24].
- The catalogue of all lunar samples, including reference numbers, written descriptions and collection site, along with photographs taken in the lunar curatorial facility, can be found online [25], along with the catalogue of Apollo surface and orbital photograph collections [26].
- The Lunar and Planetary Institute provides access to all abstracts submitted to all of the yearly Lunar and Planetary Science Conferences and special topics conferences they have sponsored since the very first in 1970 [27] (Fig. 1.2).

Fig. 1.2 EVA at the Apollo 12 landing site. Apollo 12 astronaut Alan Bean filling sample tube. Astronaut Pete Conrad can be seen reflected in his visor. (Image Credit: NASA/MSFC History Office)

- The NASA Regional Planetary Imaging Facilities [28], in particular the Lunar and Planetary Institute in Clearlake, Texas [29] and the USGS Astrogeology Branch in Flagstaff, Arizona [30] have extensive collections of hard copy and online materials. The LPI collection, much of it online, includes Apollo journal articles and books, photographs, crater and prominent feature databases, and maps. The USGS Astrogeology Branch, the site closest to where the Apollo astronauts did most of their training, has the now well-documented and only existing collection of Inter-agency reports on astronaut training, simulation, and planning documents, as well the Apollo map catalog. One special feature of the USGS facility is the set of, now online, 16 mm movies for training the astronauts in field techniques [31].

One of the most useful online sources is the Apollo Surface Journal [17], which I used extensively to perform the analyses of Apollo surface activities described in the book. This second-by-second voice transcription with annotated time-stamped links for all sampling and photography activity allows analysis of astronaut's actual performance in terms of capability for distance on foot, documentation and sampling of field stations, and manual operation of tools and instruments, all as a function of time. The application of these analyses as 'lessons learned' for planning the next generation of field science activities on the Moon and elsewhere are considered here as well.

1.2 Representative Overviews

A broad cross-section of those who played important roles in the Apollo program have had oral history interviews [22]. I would like to quote the views of four, most of whom I have known, who played critical, different, and complementary roles in the selection and exploration of the Apollo landings sites. These included Farouk El Baz, Harrison Schmitt, Lee Silver, and Jim Head. I will refer to these interview segments throughout this book.

Farouk El Baz played an important role, as Supervisor of Lunar Science Planning at BellComm, in organizing and identifying principal features on Lunar Orbiter photographs, frame by frame, to provide selection criteria for lunar landing sites. He emphasized the need for the scientists, planners, and astronauts to develop a 'regional view' to put the samples into context. I met him after the last Apollo mission, when he had become the directory of the Air and Space Museum Center for Earth and Planetary Studies. Here is how he described the process (Figs. 1.3 and 1.4):

Fig. 1.3 Farouk El Baz discussing landing sites. (Image Credit: NASA HQ History Office)

Fig. 1.4 Farouk El Baz reviewing maps with Ronald Evans and Robert Overmyer. (Image Credit: NASA HQ History Office)

…we had three-by-five cards… I'd write the mission number, the frame number, and then write some notes on what that picture contained, what features, one by one. Went through them all, couple thousand of them, and organized them, and then when I finished I asked myself just a very simple question. Why does NASA need geologists? It is to select landing sites on the Moon, so if we want to go and land next to every single type of lunar surface feature, which means that we would sample every kind of lunar material or rock, if all of these features have different rocks, where do we go? I took my list, the card file, one by one, went through them all—this is here, this is here, this one here. I came up at the very end with a list of 16 places on the Moon…

…when we're asked something we answer it for the good of the program, regardless of whose idea it was, who in NASA Headquarters would like it, and who would not like it. It is irrelevant. It is for the good of the program, period…

…Rocco Petrone was the dynamo behind Apollo. The Group_for_Lunar Exploration planning was under Wilmot Hess. GLEP had to consider all of the scientific aspects of the Apollo missions: the chemistry, the rocks, the geology. looking at the Earth from the Moon, the solar/laser reflectors, the seismometers. All of the measurements, the heat flow measurement, all the experimentation. all the experimentation, all the science that will be done on the Moon's surface…

...For us scientists, the issue was is it going to be right? Are we right in thinking that this dark area is made of basalt and volcanic rock, and this is the way it is and that was what it means about the history of the Moon or not? Is it going to be really safe for the guys to land or not? There were all kinds... We were not taken by the enormity of the event at the time, because we were too heavily involved and too thoroughly obsessed by the details of how do we do it and how do we do it right, so we never really reflected at the time...

Further describing the astronauts, their training, and creating a 'regional context', and looking at 'the shapes of things', he says:

...[the] vast majority of astronauts weren't interest in geology early on.... We had been giving the astronauts, or were planning to give the astronauts, locations on the surface of the Moon under the orbits that were called landmarks. they gave them something like seven: if they get three or four, they're fine.

When [we] started using conventional shapes and colors (like looking at clouds), and one or two leaders became good at it and began showing off, got others involved (very competitive) and they became active participants... The competition has begun in knowledge of the Moon rather than on flying the spacecraft. These guys are going to be competing about how much science they get, they bring back.... I think it was a turning point for the whole program, it really was. It was the initiation of competition in scientific knowledge about the Moon...

...their schedule did not include any of this at all. The astronauts would invite me to go have dinner... Jim [James A.] Lovell [Jr.] actually supported this whole notion. He had curiosity of his own. When I began to explain things to the whole crew, he was the one that was most lively. You could see it in his eyes, his thinking. Calling things things, and calling them by name, and getting interested in it...

We had emphasized like two or three different things. Number one, what is the relationship between what your buddies are going to collect on the ground to the big picture. That was an important component of all of them. Look at that site, tell me that it represents what? Does it represent the whole section of the Moon going like this, or it's a weirdo that is a unique setting that we cannot generalize from it around? So what's the relationship between where the crew members landed and the regional setting? What's the big picture here? Whatever they collect, can we generalize about that? That became the relationship between the landing site where the lander is going to be and the rest of the Moon.

Number two, we have questions about the area that you're going to fly over, and we need you to answer these questions. We will tell you what they are. We'll train you to some of this and tell you what is it that we need to get out of here.

Number three, you are not an instrument. You are a human being with a brain that can do us a lot of good. There are many things in the places where you're going to fly that we have no idea. None of us flew there, we have no idea what they are. If you really look with enough curiosity and enough intelligent vision, you can add a hell of a lot to our knowledge of the Moon. These were three different components that all of them had to be put together in the mission time for the plan… The places, the thing that we worked on before the mission—and we planned it right—are what are the things that we know that we want him to do, then let's look at that landing site with great care, then you can tell us the relationship between the landing site and the rest of it. Three, you're on your own, man. You can do us a lot of good or you can add zilch. It's up to you. It depends on how you look, whether you just look, or you really see and tell us what you see. Tell us in useful words.

According to El Baz, the astronauts were trained in the use of specific terminology as well. And they were trained, flying T38's, to follow a flight plan that used navigation aids (placed) at known location and signature called VORTACS analogous to utilizing visual clues for landmarks on lunar photography. These were followed by debriefings to assess performance. El Baz continues:

NASA was very clever in dividing these debriefings into levels. Deke Slayton and the crew talked about how the crew worked together. If there were any problems between the commander and any one of them, what was that? How did it develop? Where did it end? How could we have avoided them? Then the management debriefing. The three astronauts sit like that and the whole management slew of NASA Headquarters and JSC and you name it, whatever management is, sits and the crew explains the mission and tells the management what they think. The management asks questions and they would answer. That ends. This is one session. Then we have several days. That was a concession from Deke, that we have the crew and the scientists. We can go over step by step on everything. You said this was finer, what do you mean finer? Was it the fine-grained or maybe it was covered by—exactly what finer-grained means to you. Asking, —Finer-grained than what?|| so they can get every conceivable squeeze out of their observations. That could last like three days. It's a little more informal…

El Baz also emphasized the importance of training on physical analogs, such as the East Flagstaff Volcanic Field, filled with cinder cones similar to ones in Marius Hills site planned for Apollo 18. And, he indicated that the assumption was that the geologist astronaut, Jack Schmitt, would be available for the later, more advanced missions later in the J series. He recognized:

the need to 'motivate people to go… beyond what they think is possible', a viewpoint he indicates came from Chief Administrator Jim Webb, via Rocco Petrone.

Lee Silver was a CalTech professor who became, on the recommendation of Gene Shoemaker and Apollo 17 astronaut and former Silver student Harrison [Jack] Schmitt, the primary instructor for the Apollo lander crews. Jack made the first approach to Silver, but shoemaker, chair of division, helped him make decision This training resulted in the extraordinary scientific value of the J missions. Silver in effect invented the methodology for geological field work for remote exploration. Silver saw the Apollo program as 'an incredibly complex organism of many talents, many people'. Silver expressed his views as part of the oral history interview (Fig. 1.5):

Shoemaker made public critical statements which infuriated high level [people] in Washington and [made] a lot of people who knew him well enough…respect him.

Fig. 1.5 Lee Silver points out salient feature to Charles Duke and John Young during field training. (Image Credit: NASA and USGS)

> We'd had an incredible Apollo 11 experiment, that proved we could get men to the Moon and bring them back. Then in the fall we had the Apollo 12 experiment that proved that we could not only get to the Moon, but we could pinpoint our location.

> You had to know the astronauts to know what approach would convince them Lee could help them, and for that [you] needed to convince team leader, the guy in 'command'… extremely competitive…But in terms of energy, intellect, and determination, you can't beat the astronaut crews on the average. The selection process really was superb [and resulted in the selection of] extraordinary talents.

Silver's approach, to create 'useful and exciting exercises', for which they used GROVER, the USGS rover (analogous to the lunar rover) and which involved training those who would be in the geology 'backroom' during the missions as well as the astronauts themselves, was:

> Tell me what you see, because you've got 360-degree vision, and now we want to practice articulating the hard stuff you see. Now you're going to do documentation. You've got first-rate cameras. You've got all kinds of other things. But I'm in [the] science back room, and I'm trying to see how well your landing site suits the needs that had been projected as the targets of the mission.

> We're coming down the wash, and the wash was about three times as wide as this room, and there's banks going up five, six, seven, eight feet. The wash itself is filled with dry sand, but the banks are exposures of rocks, and that's what I wanted them focusing on: what are you seeing? And some of it was very straightforward: color, shape, texture, all the other things that have to go…

> Every mission was different because we had retrieved samples which were being analyzed up to and through the time of the ongoing missions, and we were changing our understanding…

> The Apollo 14 crews did not have the right attitude, did not learn enough about their mission, had the burden of not having the best possible pre-flight photography, and they weren't ready. They got lost.… Dave Scott was a gem…

> I worked on a team called a Traverse Planning Panel [Lunar Surface Traverse Planning Team]. Their mission wasn't just to go up and pick up samples. They had to deploy very sophisticated instrumentation. They had, if necessary, to make corrections or repairs… [They were] in-trained to look around, take big picture before focus on details… [The] later landing sites grander vistas, more three-dimensionality.… [Some field training was done at} the Nevada test site for atomic explosions,

where we were studying craters, and the other was to a place on Navajo reservation called Buell Park... Hadley's Rille required finding an analogue.

Well, we had an EVA team, and we had another team which was called the tiger team. The tiger team worked when the EVA team, went to bed and ate, and the tiger team stayed up and tried to digest what had happened and how it bears on the following, on the rest of the experiments, and what have you.

[The astronauts needed to see the] flanks of mountains... you can find what's come down [from above] from the flanks of the mountains. So I took them up into the high San Juans to show them what I'll call talus slopes, debris flows, all these other things.... [The San Gabriel Mountains were used as] as analogues for anorthositic rocks...

Some of the geochemists, including Harold Urey, were convinced that the geologists were biasing the astronauts, that is, they were building in preconceptions, notions that the geologists had, and that this would, in fact, detract from the ability to perform the best experiments on the lunar surface.

Silver commented on the nature of communication, and the 'chain of command':

They never commented to me; they commented to the crew commander, because that's where the power and the action was... [the] full-out communication system with the CapCom [was] working, and we worked it as if we were going to be in the science back room.

Silver comments on the value of debriefing to apply lessons learned from the training exercises and the resulting improvements:

The word "debrief" is so important in Apollo preparation and what have you. You have to go through what you've done, and you have to know what you did well, and you have to know what you did poorly, and you have to know what you would work on to make it better... They did better than that. They found a rock about that high [Silver gestures]. They rolled it over and brought up the shielded side. They sampled the unshielded side. They sampled the shielded side. They sampled the soil underneath it. You understand, they revised the experiment and did it better....

I will say this very carefully: they did a beautiful job of sampling.

Probably throughout the entire string of Apollo missions, we had never directly sampled a rock that was still in place, what we would call bedrock. And these guys knew that one possibility was that they would find that in the rille. And they went over, and they sampled, and they took their pictures. There was a very important protocol, which every crew had to develop a discipline in, what you did to get a sample that you could use then for various kinds of science. They did…

Silver comments that many of the pre-Apollo landing assumptions that the lunar regolith was homogeneous on a global scale were wrong, including Harold Urey's view that only one sample would be needed, implying global homogeneity in the nature of the regolith, and Tommy Gold's view that the surface was a homogenized layer of fine dust:

I have tell you, seeing the truth is only as good as your eyes are and as good as your brains are, and nobody has absolute control on the truth. But there's usually, amongst intelligent people with good eyesight and a good brain, a convergence. Tommy Gold had his blinders on because he'd already publicly stated his positions, and he could not converge. But after Apollo 11 he dropped out. By the time we got to Apollo 15, we were able to see rocks and bring home samples of rocks. We had done this at 12, we had done this at 14, and we would have done it with 13 if it had just made it… So Apollo 15 sort of, if I can use the phrase, it was apotheosis of all the things we'd been planning to do, and we did so many other outstanding things in subsequent missions. But it was the coming together of developing the technical capabilities, preparing men to be explorers, as well as many, many other things, and then, I'm going to say, felicitous going. But to get down to the last analysis, it's the man, and it's the man who knows what he's there for, who is driven to do it as best he can, who has the motivation, the talent, the intelligence, the education to do it, who'll get us the most per unit of time—I won't say per dollar. [Apollo] 16 now, the models we had used for telling these guys, we now had better photography, we had geologists who were spending a lot of time preparing for this site, are not very good.

And Silver expressed views that the Apollo J mission contributions to science were major:

Apollo was now down to doing science. That didn't have the same Washington value as doing other things there. …the engineering performance was outstanding, really outstanding, and many of the flight controllers in the Flight Directorate were engineers, and they were the people, the first people to do an all-cost systems engineering, and that was a great invention. I think it was Apollo who brought that on… The human performance was extraordinary, and I've already given you individual things, but what I want to tell you is about the guys who work all night, year after year, to make sure that their responsibilities are up to snuff.

1 Overview: The Apollo Lunar Surface Expeditionary Approach

The best view of Apollo, we've already touched on. That's the ability for people, a spectrum of people, from the brightest to the—and the spectrum of people includes talents at various levels and in various disciplines and arts and what have you, for a spectrum of people to do extraordinary things.

Silver also commented that, as the automated flight support systems failed, the missions wouldn't have been successful without the manual override.

Jim Head, now a well known professor at Brown University, is largely responsible for the University's leadership in planetary geology. He began his career at Bell Comm, as a systems engineer supporting the Apollo landing planning. He describes this experience below:

What BellComm did, and, for me personally, it was just amazing to parachute into a bunch of people, engineers, scientists from all disciplines, who were looking at these things from a systems engineering standpoint, whether they knew it or not, and the real systems engineers, which were the people from Bell Labs and the NASA people, were integrating it together in such a way that it worked…

We need to have six backup landing sites to present to Congress next week [for] after Apollo 11." It was, like, my God, what, "Sure." I was too dumb to say, "I don't know what I'm doing." I said, "Okay." So it was a lot of on-the-job training. It reminded me of the Berlitz [International] method of teaching languages, like total immersion.

So I did end up soon thereafter getting into the geological traverses, mostly the surface geology, into background studies for landing site selection, for doing geological analyses and mapping to help support the USGS people.

Jim Head comments on the challenges of planning safe landings and 'doing science' when the details (including being limited by 40 m resolution images in the early days of planning the landing) are unknown (Fig. 1.6):

A lot of science is like that, where you're really trying to understand the detailed relationships between things. But on the other hand, if you step back, in the final analysis what you're really trying to do, broadly speaking, is put all this together in understanding how the Earth works, where it's been, where it is now, and where it's going. So it's a system, big-time.

The questions that were being asked at the beginning of the Apollo Program were very, very basic. They were, had the Moon ever undergone extensive melting or not, what is the origin of the Moon, what makes up the dark materials versus the light materials.

Fig. 1.6 Jim Head at NASA 50th Anniversary celebration at Langley Research Center in 2008. (Image Credit: NASA LRC)

The engineers were all trying to find areas that were safe to land, which was their job. But that didn't mean that it was the best place to land; it just meant that they felt that you could safely land there. So almost immediately there was introduced two parallel tracks; where do we want to land because of science, and where do we want to land because of engineering and crew safety? As you might always predict to be the case, they were totally orthogonal…

As to what in particular was BellComm's role, Head says:

BellComm was in an unusual position in a lot of ways. This is my view, now. In many ways it was detested by other parts of the program because the people at the Johnson Space Center [formerly Manned Spacecraft Center, Houston, Texas] thought they were running the program, and indeed they were in many ways. The people at U.S. Geological Survey thought they were doing the geology, which they were in many ways. And the people at Headquarters felt that they needed some kind of organization to make sure that all this was coming together, because they weren't just administering. They wanted to be sure… So that's where BellComm came in, to do the systems people, because they saw you as an interloper in some ways, you know, just initially, a Headquarters spy…

So if you had a meeting, and I could only remember just a couple of times when any kind of interpersonal rivalry emerged in any of these meetings. And when it did, it was like you'd pooped in the punch bowl or something. It was, like, everything stopped, everybody, you know, "We're going to the Moon next week. I mean, come on, can't you guys do this somewhere else?" And it was incredible. So for me it was a real demonstration that if you agree upon a goal, people can really work together to get it to happen.

Head described the importance of the five Lunar Orbiter missions:

…automated orbiters, that acquired images for the Moon for site characterization, certification, and so on, and as they got more and more data, they were able to turn these to things other than yet another set of flat areas with which you could count craters on, and they began to look at advanced science sites… So the selection of those advanced science sites was really going to determine your menu for future sites, and there were quite a number of them…

And Head further describes the importance of 'lessons learned' from previous landings:

One of the smartest decision, I think…was made by Bob [Robert R.] Gilruth…to go to the Apollo 12 site, what became the Apollo 12 site, which was the Surveyor III landing site…

…Over Hadley mountains looking up…they had to pitch over and land over this side of the rille, and they landed very close to the landing site. The same with 17, it was a box canyon. So increasing confidence on the part of the engineers was mapped out into more and more complex landing sites, which increased the science big-time. There was a parallel thing which you allude to later on in some of these questions, that the relationships between the scientist and the engineers and flight directors, due to a lot of very insightful opportunities that people had and took advantage of, that changed over time, big-time. The interaction increased, the understanding increased, and the science benefited immensely from that.

So by Apollo 14, the engineers and flight directors were pushing to go to places that the more conservative engineers didn't want to do, and they won. That's why we went to Apollo 17… You made your estimate and then you add what we call now in mission planning 'reserves'. The easiest thing is not to do that, to be safe. You say, "No, we can't land there. Just go to the other place," and it's a safe landing. You're a hero, you know? But these people ultimately knew, based on our interactions, how important it was, and so they went to bat for you. It was really, really quite amazing. But slowly people began to realize that, okay, if we're going to continue to do this, that we have to be able to figure out a way to optimize the mission. Because clearly any of those kinds of people also wanted to do the absolute best job of whatever the mission was.

Head also praised Dave Scott as

> Amazing. He really wanted to optimize the geology, and he did put pressure on to do certain things. Like, that's the only mission that had the stand-up EVA, because he wanted when he first got there to go look…

And Head describes some of the limitations that were overcome:

> There was a concern about them getting into craters and not being able to get out. So it wasn't so much geological training as actually real terrain training, basic geological things, but also having somebody go down into the crater they can't get out of and then having—they had kind of like a tether-like thing that they carried. I don't think they carried it past 12 because the problems kind of went away, I think. But, you know, testing on that, like, okay, throwing it over and bringing them back up and so on.

> So it was a combination of real-time sort of engineering-type training, geo training and geology. Then it just more and more, more and more details, like, yes, tons of time practicing as the J mission.

And Head comments on the nature of the training as:

> …more specific kind of training, not just geological training, but traverse training, as it got closer to the missions. But in the intermediate time we'd take the crews out, once they were named, to analog-type areas.

> Geology's kind of low on the list for a variety of reasons, mission success and so on. So what he instituted was us coming in in the evening, which was really their time… then after that, we'd adjourn to the briefing room next door and we'd spend an hour or two on the traverses and whatever the geology, and it was great fun.

> [The experience was] totally adrenaline, because it was a limited time on the surface of the Moon. You just end up not knowing how long they were actually going to be there. You never knew when something was going to happen. The only thing you could be sure was that something was going to happen that you didn't plan on. So it was a total adrenaline rush, and there actually wasn't a lot of time for reflection of that kind, you know what I mean, in that sense. But it still was, again, part of that teamwork thing. It wasn't like "I did this." It was like "It's really working."

And there was a lot of emphasis on 'focusing on what you are seeing':

Okay, what are we going to do here? Do we go all the way along there, or do we spend time where we are now to optimize what we're looking at? There are some boulders here, but we don't see many down there." So the decision was to really focus on what you're seeing, what you've got here. And that's where they found the green glass, actually. So those kind of calls...

[For] Apollo 17, there was a gravimeter measurement made at the edge of the southern mountain range there, the South Massif. Its value was so different between what was taken around the lunar module and there, that they really wanted to try get one in between. So, the call was, okay, change the traverse, that you stop on the way back to deploy and get a gravimeter measurement. That's one I remember distinctly because it's so clear. You need to establish the gradient. Is it linear? Does it fall off or whatever? It's also one we did in one of the simulations. That's one I introduced. I introduced some readings which would force you to have to make that decision. So I put them in to force the people in the back room to make a decision, because it's so obvious something was going on with those readings. And that's essentially what happened...

[In the] Group for Lunar Exploration Planning...at some level there was a decision made about, okay, heat-flow experiment has the highest priority, field geology is next, and so on is next, something like that. Commonly they'd be broken down into elements, to deploy it, or do this or that or the other thing. Then you'd argue about that, you know, about which had the highest priority and [what] the mission rules were. So there was a group of people who kept the mission rules, and that's where the priorities get mapped out into. It's like anything; you have a set of rules like that, like default things. If this problem arises, then the highest priority is you go back. You get the heat-flow probe and before you take off and do more geology-type things. And they just got argued out as a function of time.

[The] Group for Lunar Exploration Planning and the USGS and JSC and BellComm, etc., all had input into this whole thing, but ultimately a set of recommendations were made for the site selection to the ASSB, the Apollo Site Selection Board, which was chaired by the Director of the Apollo Program and had representatives from the major centers and they made the final decision.

When asked about the site selection, Head said:

Well, mostly the Lunar Orbiter photographs were really important. The engineers did a whole host of analyses related to roughness, measurements based on crater frequency distribution and a lot of other things.

...It wasn't until we were able to get them to agree to a plane change that we could go to sites like [Apollo] 15 and 17 and 16 and so on.

But at the same time you could measure things remotely with a telescope, and that information started to come into play as spots began to be characterized on the surface and differences were seen.

Head commented on how the astronauts were trained:

We used [the lunar orbiter photographs] in the training, totally, yes, always, yes. Then the traverse maps, the ones that they took with them to the Moon, were those maps with the traverses on top. So this is what they carried around on the surface [cronopaque].

Once we got the traverses down, we'd work with them on the cuff checklists to put all the right things in and be sure that—yes, here are some examples here, just be sure that—here's stuff like investigate talus for the commander and lunar pilot, what you're supposed to be doing, a little bit about the kinds of trenches to dig, what to take, and so on. We'd also always put little surprises in there from place to place.

[There were] good relationships among the crew and the crew training people... The crews always [had] a presence of mind that just, at the time, that was just normal, because you're all in this kind of together. They're working. They happened to be the ones on the Moon... Like, we called Jack Schmitt up, I think, the night before launch in crew quarters. He was so hyper on the geology, we said, "Don't forget to enjoy the trip," you know what I mean. It was, like, it was just a lot of fun because the people were so tied up and things. But at the same time they would do things.

...to document the rocks, [we took] a stereo pair and a panorama and [then a] "before" and... an "after."

Head Commented on the post-trip briefings:

Yes, the question of being involved in the post-flight activities, we were involved in a couple of ways in the debriefings of the crews when they came back. This became increasingly important and increasingly personal as the missions went on and the group became really more focused and smaller in some ways. For example, Apollo 11 and 12 debriefings were a lot of people because, well, of course, it was a big deal, so everybody wanted to be there, but also everybody wanted to learn from it, too. So they were more formal kind of affairs that there were a hundred people there, something like that, maybe more. I can't remember now. But probably most of the real debriefing took place on a more personal level with the people who were key in the next mission.

But with time they became more personal in the sense that, I mean, I can remember participating—well, eventually what we'd end up doing, and I can't remember whether it was the formal debriefing or the informal debriefing, but we'd take all the tapes, the photos and stuff and go to some room in the basement of mission control and just sit there for a day and go through these things. That was just amazing, because for the first time you would see the astronauts responding to them seeing themselves on the Moon from the films, the TV cameras and things like that, which, of course, was totally different for them. So they could see themselves interacting, and, of course, they were pretty relaxed at this point, and it was hysterical. That was one of the funnest times, because we'd just sit back and go through it with a set of really good questions to try to optimize the next mission. But it'd just be a lot of fun, too. It was, like, they'd be calling each other klutzes when one kicked over this or that or something like that, and it was just a lot of fun…So that was really important, because actually for us it gave a much better picture for what they really felt they were capable of doing and what was easy, what wasn't easy.*

And Head commented on the simulations:

There were a number of different kinds of simulations. There was the kind we did in the field where they'd be out there, they'd be doing their thing, and we'd be pretending like we were on the Moon. They were great.

But there were full-up simulations where all of mission control was simulating the mission. That early on introduced the problem of actually how you do that. So you couldn't early on do that from the remote site. You couldn't say, "Okay, we're going to be out in Arizona and hooked in with mission control," because there's an incredible communications mess. Eventually one was able to do that a little bit from time to time, for example, from the rock pile at the Cape, for example, where they were able to kind of, like, get in the suits, do their whole thing, and so on. But mostly earlier on it was just simply exercising the systems.

…we'd have a simulation. Okay, it's going to be August 12th. We're going to simulate EVA 1 of Apollo 15. So I would have to go to Houston and get together with the simulation supervisor. These people were incredible. They were the most feared people. So the sim sups, it's like they're God. So they're going to control what this thing's going to look [like]. They were actually a bunch of really good people. But they would say, "Okay, what we want to do here is we want to exercise a number of different kinds of systems," because I'm worried about this system, that system, maybe the suit or the rover or the whatever, whatever.

We'd get together beforehand, myself and a guy named Hiram [G.] Baxter, who was an [Philco] engineer down there, and we would be the two pretend astronauts. It was one of the most unbelievable experiences I've ever had, because you'd, first of all, be in on the whole thing. So the sim sup would, say, "We need to exercise this time these five systems. So you need to, Jim, figure out a way we can exercise those systems."

So, you know, I had to, for example, get a suit leak that wasn't terminal but was slow. So I had to think of a way in which, you know, I'd be going, like, "Oh, Houston, you know, I see this really interesting rock over here. It looks like a multiple breccia," something that would entertain the people in the back room and get them excited, the geology dudes, but at the same time rush over there, trip on the rocks, fall over, say, "Oh, I'm okay. No problem," and then drop my hand, and they'd introduce the suit leak. So it had to be something integrated that made sense. So I'd be going, "Oh, I see a really interesting block field coming over. I'm going to try to get near the crater," and then I'd pretend I'm driving towards the crater. "Oh, it really looks neat. I can see these big boulders here, there, and everything." Then the wheel would fall off or something like that, and you'd do the wheel fall-off thing.

So we really worked together to pretend like you're on the Moon and make it something where the people in the science back room were being exercised, as well as then make it an integrated thing where the flight controllers and the whole flight operations setup would get realistic geologically, so to speak, realistic problems that would arise out of the surface operations, if you get the drift. So that was a lot of fun.

When asked if the findings met their expectation, Head said;

Well, it's sort of back to the 'We're making this as we go along.' I think so much was new that it would be hard for it not to meet your expectations. I mean, almost everything we were learning was new, and you didn't want to get complacent about that. There were people who would be satisfied with just landing at some point and picking up rocks. You have to appreciate that a significant number of the people who had worked on extraterrestrial rocks, they were meteorites, and nobody had a clue where they came from in context, geological context. They'd fall out of the sky, you know. So the idea of spending time to figure out exactly what—they'd done profitable analysis on those kinds of rocks for years. So they weren't so concerned about geological context. But the geologists really were, and probably went a little overboard on that, too.

I think the best example of [unmet expectations] is Apollo 16, where we went to a site which we thought we knew what was going on, we didn't know what was going on, we found out something completely different than we'd anticipated, and lunar science is the better for it, totally, you know what I mean, from that point on, because, in fact, it wasn't what we thought was there, and that's what exploration is all about. So, in that case, there's an example of something where one might be disappointed we didn't find what we were looking for, but they'd miss the whole point of what exploration was all about.

As to the impact of Apollo on the rest of their lives, Head says:

> *I think that was a privilege and an accomplishment at the same time, from my point of view, anyway… It's also made me appreciate the failings of working together when goals aren't well articulated and agreed on. So I think Apollo is a real tribute to human capability…*

Harrison Schmitt, the only professionally trained geologist among the astronauts, and Gene Cernan, were the landing crew for the Apollo 17 mission. Prior to that time, he had been one of the astronaut trainers. While on the Moon, Schmitt collected troctolite 76535, providing the most compelling geological evidence that the Moon once had an active magnetic field. Since that time, he has been a US Congressman from New Mexico and a university professor. From the time of his earliest involvement in the Apollo program, Jack has continued to play an active role in lunar geology. I've seen him at many of the lunar meetings I have attended, always commenting on and asking some of the most scientifically interesting questions (Fig. 1.7).

Fig. 1.7 Harrison Schmitt training in the Lunar Module simulator. (Image Credit: NASA KSC)

This is what Jack had to say about the Apollo astronaut corps and their training:

I don't think it's gotten nearly enough credit: what Al Shepard and Deke Slayton decided to do with the astronauts who weren't directly involved in flights. Weren't assigned to flights. And they were—maybe partly because, "What else do we do with them?" but partly because I think their insight was correct. In that having astronauts who were going to be at the tip of the spear, part of the review teams, part of the Critical Design Reviews, getting very familiar with certain aspects of the spacecraft, the rockets, and the planning and operations of Apollo, was extremely important in making it real to everybody else that people are actually going to use these things that we're building.

My assignment was to oversee the development of the scientific experiments for Apollo; the so-called Apollo Lunar Surface Experiment Package [ALSEP].

If you saw a problem, you began to work the problem; and nobody ever resented the fact that you started to work a problem, whether it had to do with something going on in real time in a flight or whether it was something that you anticipated might be a problem later on. That whole environment, I think, sort of disappeared for a while. And particularly as we led up to the Challenger [51-L] accident. It's been my strong impression from things that have been said that having astronauts farmed out around the country, watching over hardware, really wasn't taking place in that time. And the best sign of that is that the—apparently none of the astronauts were aware of the anomalies in the solid rocket booster that had been identified by the Marshall Space Flight Center [Huntsville, Alabama] during some of the earlier pre-Challenger launches.... Well, by then it was too late. Because they had received guidance (again from astronauts) some time before that—"to give us something to do on the surface" (and these were from non-geological astronauts who apparently felt that they were going to be bored on the surface of the Moon, which is a little bit hard to believe). So, when Anders and I got up there, we had to do an awful lot of work to get this thing down to the point of where it was not going to occupy all the time that the astronauts would have on the lunar surface.

There is another place where Eugene Shoemaker made a great contribution. His initial thought (and one which we had had together, while I still worked at the US Geological Survey) was a concept called the surveying staff, so that we could automate not only the photography but the positions and the orientation of the cameras and everything could all be done in one staff that—and all this data would be automatically recorded. You'd have a range finder, and that would go back on the data link. And so, the astronaut would be free to do the things human beings do best, and that is observe and think about what they're seeing. Unfortunately, that particular concept fell apart, not because NASA didn't try to do it but because they were—as I understand it, were forced into a small business set-aside contract. And whereas a company—a large company like Kodak, say, could have done this…

And Astronaut Schmitt had very useful comments about the tools they used to explore the lunar surface:

Just very simple geological tools, which have worked well through several hundred years and probably will continue to work several hundred years in the future. But still, it requires a great deal of human action. And the whole idea then—as well as in the future—of putting human beings on the Moon is to minimize the number of routine tasks that they will have to do so you can maximize their contribution for what human beings uniquely do; and that is, look, manipulate in detail, and also think about what they're doing. And what they see. But still we did the best we could. Shoemaker's and his group, I think, greatest contribution to the science of field geology was from a tool point of view was the what's called the gnomon. It was a device that we would set in each of the pictures (particularly stereo pairs)—a device that had a gimbaled vertical rod that would give us the vertical orientation of the gravity field. It was of a known length (40 cm, if I remember correctly), had a gray scale on the rod, on—it was a three-legged device. Its shadow—the shadow of that vertical rod—would give us azimuth, because we knew where the Sun was, of course, and we could always tell what that was from the shadow on the ground. And there was a color scale as well on one of the legs that can be used in order to print the color pictures more accurately.

Al Shepard bought the first Hasselblad in a shop in Florida just before his suborbital flight. NASA was not going to supply him with a camera which shows you every once in a while you come into little glitches of thinking… And then after that Viktor Hasselblad, the owner of Hasselblad in Sweden, as I understand it, volunteered to adapt it in various ways for more easy use by the astronauts. He introduced for the first time a motorized winding system, a trigger system so that we could operate it on the surface.

…There's no question that [Gene Cernan] could handle the hammer better… We never got the design of the handle so it would universally be compatible for all the astronauts. And Gene just had a very big hand; and he could…grip it better and not get as fatigued using it. (I used it every once in a while.) But on the other hand, for…geological purposes, I found the scoop much more versatile. I could pick up rocks with it—small rocks—and I…could also dig trenches and do all sorts of other things with it.

Again, Harrison Schmitt commented on the selection of landing sites, indicating everyone, scientists and engineers, realized that the primary objective of Apollo was to land safely and return safely to Earth:

> *The Lunar Orbiter…was flown specifically to try to identify landing sites, particularly within what was called 'the Apollo belt of the Moon', the equatorial belt of the Moon, that were as smooth as possible… And a number of landing sites were evaluated, in great detail, in various ways, in order to select potential sites within that belt; and they were sites roughly that were a 1-day slip apart that—based on lighting considerations. We always wanted to land with the Sun behind us, at an angle lower than the glide slope—than the slope we were on—in order to come down towards the lunar surface so you could see shadows…*

And Harrison Schmitt also indicate that the ALSEP surface instrument package evolved as resources, including planning time, become more scarce:

> *Because of the intensity of efforts [in planning for Apollo 8], NASA had never really sat down and worked out a total flight plan for the Apollo 11 mission. We were taking it one step at a time. And in the process of putting that timeline together, it became clear that the chances of us ever flying a full-up ALSEP were very small. It was over 300 lbs. We still didn't know how much margin we were going to have in fuel for landing on the Moon. It was a clear candidate to be offloaded at the last minute in order to increase, by a number of seconds, the amount of time we had of fuel. And so, I made a proposal that we begin to look at an alternative, which would be a slimmed-down version of an ALSEP, something that would not have a radioisotopic thermoelectric generator, because that was quite heavy, but would have solar cells to power the experiments through 1-day cycles on the Moon. And I began to make phone calls around the country, asking my geological colleagues and geophysical colleagues, "If you thought you would never have more than one mission, what were the priority of experiments that you would want to have on the surface of the Moon?" And it became very clear, very soon, that everybody (geophysicists, geologists) were saying, "We need a seismometer. We need to understand as much as we possibly can about the seismic environment of the interior…and we want a corner cube reflector.… We did design that little package of experiments…what Neil [A. Armstrong] deployed was a solar cell-powered seismometer and a corner cube reflector that is still up there and still operating.*

Harrison Schmitt was also organizing the training program for the post-Apollo 11 missions:

> *And at the same time [as planning for the Apollo 8 mission], I…was having to try to organize a brand-new training program for the Apollo mission, particularly the post-Apollo 11 missions. That had started, really, late in 1966 when I, after having thought about what was going on in the science training of the astronauts, having participated in some of the lectures, participated in—or had attended some of the lectures (not given them), participated in several of the field trips that the astronauts had been*

given as grouped—as a group, it seemed to me that we were really off track. Number one, we were boring the astronauts. They were not interesting. The intent, apparently, was of the people then in charge to create—in their words, "create astronauts with Master's degrees in geology." Well, that's not what we needed. What we needed were focused, relatively narrowly trained field geologists who primarily were pilots, but people who could select—who could observe and select the widest variety of rocks, could tell us about the context in which each of those samples came from, and so forth.

I [proposed] that we begin to focus our science training on an actual simulation of lunar traverses in areas on Earth where we could learn something about the kinds of problems that we would encounter on the Moon. (There are no real analogs on Earth, but we could focus on concepts and actually run short traverse exercises during this time—during the time they had for science training.) And number two, is to go out and recruit the best teachers that we could possibly find to do this, who would not only—were very good scientists but very good teachers, and who could understand what we were trying to do and not be trying to change our emphasis, but to live within that. And in so doing, I called people like Richard Johns, who was an old teacher of mine at Caltech, one of the best field geologists that ever lived. I got a hold of Bob [Robert P.] Sharp, also a Caltech professor of mine who was the best observer of detail and what he called "belly geology" that I had ever encountered. Lee [Leon T.] Silver was invited to participate. And I called a number of lecturers, particularly people I knew, from Harvard such as Jim Thompson and Jim Hayes, Gene Simmons, and others to prepare lectures that were stripped of the vocabulary of geology and used the vocabulary of general science (of physics and chemistry).

… The lecture program was reoriented much more towards what would be useful to them when they got to the Moon. And the field program, about a one a month training plan was oriented toward field simulations, where the field geolgy was learned and geological concepts were learned, but in the context of actually performing as if you were on the Moon in everything but a pressure suit. That made all the difference I think.

US Geological Survey people (under Gene Shoemaker) had already been working on for a long time. And they were really the technical core of our training program. We always depended on them to organize the logistics of the trips, to recommend the sites that we would go to, to lay out the traverses that we would want the crew to work on. They were really the hard rock core.

…We went out for 3 days and ran traverses like we talked about and then we'd go back over them so they could see what an experienced geologist would have seen and the critique was done in near real time… [The astronauts] came back very enthusiastic… They said 'Yes, this is what we want for our training program'.

> We really brought in a national program of training that I think resulted in what we have today...a first-order understanding of the Moon as a planet, and as a consequence, a first-order understanding of the early history of the Earth, and the terrestrial planets.

Harrison Schmitt commented directly on the faith upper management had in and the resulting lack of micromanagement and flexibility given to the geological experts:

> One thing that is implicit in all of this is that we were allowed to do all of this by people like George Low and Bob Gilruth and Sam Phillips and other who were the senior managers... They don't get the credit they should, particularly from the scientific community.

> [Senior managers] had all the confidence in the world that they were going to land people on the Moon and return them safely to Earth...but once we did that, we were going to have a capability to explore the Moon...to do things far more than just that, that's what the lunar module looked like. And so, the various modification programs were put in place well before Apollo 11 that were leading us towards a Block II lunar module that could be-ultimately was used on 15, 16, and 17-some kind of mobility on the lunar surface. For a while there was a competition between a lunar flyer and a lunar rover: the rover obviously won out.

> [Ed Wolfe and Bob Sutton, from the USGS in Flagstaff] were able to give Kranz a demonstration as well as a full-up briefing on what a science support room could do to enhance a mission...And Kranz bought it to his credit. ... That really ended up contributing a great deal to what we were able to do. And if nothing else it contributed to the enthusiasm of the scientist to work inside Apollo rather than outside.

> ...it's illustrative of the way Apollo worked. And if somebody had a good idea and you could convince a few other people you had a good idea, you could work it up through the system and eventually somebody would listen to you. It was never stopped. There were really no barriers to good ideas. And it worked in the planning stages, and it worked in real time during a mission. If a problem developed during a mission and somebody had a good idea on how to solve that problem, you had as much chance of getting that idea pushed forward as anybody else. And it—so much of Apollo's success depended on that.

> That's why you didn't do it for $20M—$20B. You did it because these people believed it ought to happen. And that kind of environment is just something that we see too seldom. Too often it has to be war. In this case it isn't war. It was competition, but it certainly was not war. And those people were the reason that you could get

almost anything done. There was never a paucity of ideas. Imagination was rampant, and most of it very good imagination on how to solve problems. And a group of people could get around a table, work together, and in a noncompetitive—it seemed noncompetitive, at least at the time—and the sum of the output of that table was far greater than just the individual parts that were there. It was really an exciting time to be involved. And that's why Apollo 13 was saved. That's why Apollo 11 landed at the time it did. It's really why any of the in-flight emergencies were dealt with successfully, is because people could get together and figure out how to solve the problem.

Jack had comments focused on experience with his mission, Apollo 17, with further implications for planning future human exploration missions:

Once you have the developed the infrastructure—capitalized, if you will, the infrastructure—on the backs of some crass, profit-making enterprise, then you can start to do all sorts of things. You've lowered the cost of access to space by a major factor because of these other ongoing activities.

The science [on each Apollo mission] was quite different because of the difference in the site. But still, the planning concepts and the flight concepts all were pretty much the same. So, in that regard, it was very familiar territory to go through.

…everybody had maintained and was maintaining this high level of motivation and dedication to success for the Apollo mission, even though we all knew, and had known for some time, that [Apollo 17] was the last of the Apollo missions to the Moon.

…I already knew in the general terms the kinds of different things that we could expect to work on. It was the most highly varied site of any of the Apollo sites. It was specifically picked to be that. We had three-dimensions to look at with the mountains, to sample. You had the Mare basalts in the floor and the highlands in the mountain walls. …You're trying to integrate your eyes and your hands and your mind, the database that exists in your mind, to rapidly as possible, and particularly on the Moon as rapidly as possible because you can't go back, get as much pertinent information about the origin and evolution of geologic features as you possibly can.

With Apollo, we planned the heck out of this thing right from the very beginning, the traverses and where's the most likely place to gather good information and things like that. So that was a net advantage. That really raised your efficiency of exploration to take advantage of the time available.

The primary disadvantage that almost overwhelmed everything else was the inefficiency of using the spacesuit. …The biggest challenge that we will have, whether public or private sector goes back to the Moon or goes to Mars, is to improve the efficiency of that suit. The other part of the glove that was a problem is that no matter how closely you cut your fingernails, every time you reach for something or moved in that glove, you would tend to scrape your nail against the bladder of the suit, the rubber bladder. I even wore liners, nylon liners, to reduce that, but still you would do that, and you'd gradually lift the nail off the quick. That is painful to some degree while you're working, but particularly gets painful later after you've gotten out of the suit and you then prepare for the next day.

Now, the nice thing, though, about working very hard in space is that the next day you have no sore muscles. The efficiency of the cardiovascular system in cleaning out the toxins, the lactic acid and other things that are produced, metabolic products that are produced in the muscles, is so great that there's no muscle damage. There's fatigue, but not damage. The next day it's as if you never did anything. You just start all over, and hopefully you've learned a lesson and you don't move quite as fast.

References

1. Compton, William. Where no Man has gone before: A History of NASA's Apollo lunar expeditions, (NASA SP-4214, 1989), Dover Books, 432 p., 2010.
2. Cortright, Edgar, Apollo expeditions to the Moon (NASA-SP-350, 1975), Dover Books, 386 p., 2019.
3. Orloff, Richard, Apollo by the Numbers: A Statistical Reference (NASA-SP-2000-4029, 2000), CIA Publishing, 395 p., 2013.
4. Launius, Roger, Apollo: A retrospective analysis. (NASA Monograph in Aerospace History, #3, 1994), www.militarybookshop.co.uk, 122 p., 2011.
5. Glennan, T. Keith, The Birth of NASA: The Diary of T. Keith Glennan (NASA-SP-4105, 1993), CIA Publishing, 428 p., 2013.
6. Launius, R. and J.D. Hunley, An Annotated Bibliography of the Apollo Program, DigiCat, 181 p., 2022.
7. Beattie, Don. Taking Science to the Moon: Lunar Experiments and the Apollo Program, Johns Hopkins University Press, 356 p., 2003.
8. Swift, Earl. Across the Airless Wilds: The Lunar Rover and the Triumph of the Final Moon Landings, Harper Collins, 384 p., 2021.
9. NASA JSC History Collection. October 1, 2024. https://historycollection.jsc.nasa.gov/JSCHistoryPortal/history/history_collection/uhcl.html
10. Encyclopedia Astronautica. 'Apollo'. October 1, 2024. http://astronautix.com/a/apollo.html
11. Lunar and Planetary Institute. Lunar Sourcebook. Heiken, G., Vaniman, D., French, B., Eds., 756 p., Cambridge University Press.

12. National Archives and Record Administration. 'NASA Apollo Archives'. October 1, 2024. https://search.archives.gov/search?affiliate=national-archives&page=1&query=nasa+apollo+archives&submit=&utf8=
13. Smithsonian Air and Space Museum. October 1, 2024. https://airandspace.si.edu/explore/collections/search?edan_q=apollo
14. NSSDC (GSFC). 'The Apollo Program'. October 1, 2024. https://nssdc.gsfc.nasa.gov/planetary/lunar/apollo.html
15. NASA HQ Apollo Collection. https://www.nasa.gov/wp-content/uploads/2023/02/apollo-subject-files-3.pdf?emrc=537cb0
16. Catalog of hardcopy Apollo Program Documents. https://www.nasa.gov/the-apollo-program/
17. Apollo surface journal. https://www.nasa.gov/history/alsj/
18. Apollo flight journal, https://www.nasa.gov/history/afj/
19. Apollo Geological Tool Catalog. October 1, 2024. https://www.nasa.gov/history/alsj/tools/Welcome.html
20. Lindsay, Hamish, Apollo Lunar Surface Experiment Packages, October 1, 2024. https://nasa.gov/history/alsj/hamishalsep.html
21. Lunar Rover Specification and Performance Documents 1. October 1, 2024. https://www.lpi.usra.edu/lunar/documents/NTRS/collection2/NASA_TM_X_66816.pdf
22. Lunar Rover Specification and Performance Documents 2. October 1, 2024. https://www.lpi.usra.edu/lunar/documents/NTRS/collection2/NASA_TR_R_401.pdf
23. Lunar Rover Specification and Performance Documents 3. October 1, 2024. https://www.lpi.usra.edu/lunar/documents/NTRS/collection2/NASA_TR_R_401.pdf
24. NASA JSC Oral History Collection. October 1. 2024. https://historycollection.jsc.nasa.gov/JSCHistoryPortal/history/oral_histories/participants_full.htm
25. Apollo Lunar Sample Catalog. October 1, 2024. https://www.lpi.usra.edu/lunar/samples/
26. Apollo Surface Photo Catalog. October 1, 2024. https://apolloarchive.com/apollo_gallery.html
27. LPI sponsored Meeting Abstracts. https://www.lpi.usra.edu/publications/abstrc-reports/
28. Regional Planetary Imaging Facilities Listings. October 1, 2024. https://www.lpi.usra.edu/library/rpif/locations/
29. Lunar and Planetary Institute RPIF. October 1, 2024. https://www.lpi.usra.edu/library/lpi-rpif/
30. USGS Astrogeological Branch RPIF. October 1, 2024. https://www.usgs.gov/centers/astrogeology-science-center/science/astrolik
31. Phinney, William, Science Training History of the Apollo Astronauts, NASA SP-2015-626, 2015.

2

The Challenge: Constraints and Requirements of Extreme Exploration

2.1 Remote Exploration Before Apollo

The constraints for lunar surface exploration imposed by the Apollo architecture, and, barring any major technology breakthroughs, by current architectures under consideration for deep space exploration as well, are far more demanding than those placed on any remote expeditions on Earth, in terms of cost, logistical complexity, environmental extremes requiring life support, deliverable mass, volume, power, and bandwidth, mobility, access, and dexterity on location or between locations, and available time.

How similar or dissimilar were salient characteristics of earlier remote expeditions requiring major resource investments over periods of years? What kind of transportation and equipment was utilized by these expeditions? What were the selection criteria for expedition members? What characteristics did successful expeditions have in common? What criteria should be considered in defining 'success'? Just as in the Apollo missions, these were crucial considerations.

As discussed below, planning and preparations were extensive, over periods of years or even decades. Polar expeditions, with different leadership and sponsorship and were occurring both simultaneously and sequentially. Expedition leaders and members were often involved in more than one expedition, with first expeditions acting as 'training grounds' for later expeditions. Private sponsorship was involved, forming the bulk of investments except in the case of the Lewis and Clark expedition.

2.2 Lewis and Clark Expedition, 1803–1806 (Figs. 2.1, 2.2, 2.3, and 2.4)

The exploration of the Louisiana Purchase was performed by the 'Corps of

Fig. 2.1 Lewis and Clark Expedition Paraphernalia. Top, Clockwise from upper left: sextant, compass, pocket telescope [5]. Bottom: Ink and writing tools [6]. (Image Credits: National Park Service Photos/Kephart)

2 The Challenge: Constraints and Requirements of Extreme... 33

Fig. 2.2 Detail of Paxson (1912) Lewis, Clark, and Sacajawea at Three Forks, Lobby of Montana House of Representatives [7]. (Source: Public Domain)

Discovery', 45 men led by Lewis and Clark who traveled 8000 miles on foot and horseback, from St. Louis to the mouth of the Columbia River and back [1–4]. They were commissioned by President Thomas Jefferson and financed by Congress. It took a visionary like Jefferson, **willing to embrace the risk and expense required to launch an extraordinary 'nation building' exercise** for our new country. Jefferson was greatly influenced in his vision by Captain Cook's descriptions of his voyages in the South Pacific.

What was the training and background of the men who led the expedition? What special training did these men or any of the participants, receive? All of the men had spent previous years as **military men**. The expedition leaders engaged in **strenuous efforts in preparation, studying a wide range of potentially related material and learning a wide range of potentially useful skill sets.** Jefferson had originally asked Brigadier General George Rogers

Fig. 2.3 Lewis and Clark Expedition. Top: Blunderbuss from that era [8]. Image Credit: Smithsonian, National Museum of American History. Bottom: Map of Lewis and Clarks Track along the Missouri, across the Rocky Mountains, down the Columbia River, to the Pacific Ocean. Clark, History of the Expedition under Command of Captains Lewis and Clark, to the Sources ot he Missouri thence across the Rocky Mountains and Down the River Columbia to the Pacific Ocean 1814 [9]. (Image Credit: Library of Congress)

Clark, to lead the expedition. He refused. Ultimately, recognizing Lewis' professional and organizational skill, Jefferson asked Merriweather Lewis, a college graduate, former Captain in the Virginia State Militia and his personal secretary, to lead the expedition to explore the Louisiana Purchase. Lewis began expanding his background for exploring the new land immediately, studying, as arranged by Jefferson, with well-known experts of the day, including Ellicott in map making and surveying, Patterson in celestial navigation, Barton in botany, Wister in anatomy and fossils, and Rush in medicine. Lewis studied the best available materials on the region, including journals and maps, in order to determine the best route from the northwest fronter at the time, the Ohio Valley, to the Rockies. He asked William Clark, his former military superior, to co-lead the expedition. Clark had been a member of the state militia and then the regular army, where he was commissioned lieutenant of the infantry by George Washington. When the Revolutionary War ended he returned home to manage his family's estate. Clark studied much of

Fig. 2.4 Lewis and Clark Expedition. The Great Falls, 'the grandest site sight he had ever seen according to Lewis [10]. (Credit: Montana Historical Society. Public Domain)

the same material that Lewis did, and he trained himself in the collection of scientific field data as well. The 'corps' itself consisted of up to 45 army volunteers from Camp Dubois near St. Louis, whom Clark recruited and trained. He selected men who were single, healthy, and had excellent hunting and survival skills. Also included in the 'corps' were carpenters, blacksmiths, tailors, interpreters and trackers. **In other words, unattached, skilled men with the widest range of useful trades were sought.**

What was the nature of transportation and equipment utilized by the expedition? In order to transport sufficient trade items for the Indians in keeping with Jefferson's diplomatic intentions, larger boats than originally planned were required, and, to manage the larger boats, the expedition hired contractors, or 'engages', to handle the boats. The route required a boat to transport all the men and equipment and yet light enough to manhandle under non-ideal (shallow to dry conditions).

While Clark was recruiting and training men, Lewis obtained equipment and supplies which included [3]:

- surveying instruments including compasses, quadrants, telescope, surveyor's chains, log line, reel, sextants and a chronometer
- camping supplies including oil linen for tents, blankets, steel flints, tin horns, lanterns, tools, utensils, kettles, pans, mosquito netting, fishing equipment, soap

- scales, weights, corn mills
- needles, weights, awls, thimbles, scissors, knives, vises, pliers, chisels, adzes, augers, whetstone, brads, steel tapes
- portable soup in cannisters
- tea, strong wine, salt, dry goods
- writing supplies including oilcloth, ink, writing utensils
- clothing (coats, shirts, stockings)
- weapons and ammunition
- medicines, medical books, and medical supplies
- books on botany, geography and astronomy
- maps
- gifts such as beads, knives, and tobacco, to present to Native Americans

These supplies came from a number of sources, many in Philadelphia, where Lewis was located. One item crucial to navigation was the astronomy notebook prepared by Patterson, who trained Lewis in its use prior to the expedition. The sources included [3]:

- 20 Philadelphia merchants,
- Public and War department stores at Philadelphia Arsenal,
- the armory at Harpers Ferry,
- Schuylkill Arsenal west of Philadelphia, which also acted as the shipping point to St. Louis.

What were the achievements of the expedition? By most measures, this expedition was a far-reaching success. The primary goals were discovery of the best practical route from the Mississippi to the Pacific Coast, characterization of resources, including flora and fauna, of unexplored territory, the assessment and establishment of trade with indigenous people, as well as to establish US claim of discovery (and sovereignty) over its western territories. Observations were recorded in written journals during the expedition. Journals were archived, but not systematically, so that the nature of this expedition and its results were not widely reported or studied until the twentieth century. The Corps was provisioned at the start of the expedition, for most of the journey, except for munitions and surveying equipment, not easily replaced. However, they also provisioned themselves, generating food, shelter, and replacement supplies from their surroundings. Technically, they operated autonomously, under the direction of their strict appointed leadership; **however, the expedition was very likely to have failed, through direct starvation during the harsh winters or getting hopelessly lost,** without the help of Native American tribes they met along the way. The biggest contribution of the expedition from a scientific standpoint was the major advance in understanding of the natural history of the North American continent. During the course of the expedition and upon its conclusions, many preserved specimens representing previously unseen animal species were received by the American Museum, under the direction of Charles Peale. The plant species were to be under Barton's care, though he failed to produce the anticipated natural history work. Many of these specimens were recovered and ultimately housed in the Academy of Natural Sciences. The first narrative of the expedition was created from its notebooks by Nicholas Biddle, who published the account in two volumes in 1814. Amazingly, **only one fatality occurred among members of the expedition**, and that was not due to conditions of travel, but most likely to a burst appendix. However, ailments like sunburn, dysentery, colds and flu were frequent, and snake and insect bites along with poison ivy had to be treated. Dental problems, likely not resulting from the expedition itself, were ongoing. Participants experienced outbreaks of venereal disease, pneumonia, and cholera, and the expedition encountered small pox among Native American tribes.

2.3 Stanley Trans-Africa Expedition, 1874–1877 (Figs. 2.5, 2.6, and 2.7)

The most publicized exploration of central African lakes and rivers [11–13], from the west to east coasts, a region considered to be deadly and impassible, was led by Henry Stanley, an extraordinarily ambitious man who began life in poverty and spent early years as an orphan. **Stanley, not a military man, established himself as a reporter who, through experience leading an earlier African expedition, developed a reputation as an explorer.** He labeled Africa 'the dark continent'.

Who participated in the expedition? Stanley brought along several English like-minded colleagues, including Frederick Barker and the Pocock brothers, as well as Kalulu, an English educated African he had taken on a previous trip, who eventually became has adopted son. The rest of the initial **224 people of the expedition were indigenous Africans**, most of them from Wangwara in Zanzibar, the expedition's starting point, based on their reputation for endurance and reliability. The African recruits included a small number of women and boys.

Fig. 2.5 Explorer Henry Morton Stanley with Kalulu, his gun bearer and servant on expeditions and otherwise constant companion who became his adopted child. Stanley is posing in suit he wore when he met Livingstone [14]. (Source: Public Domain)

Who sponsored the expedition? Unlike the Lewis and Clark expedition, which was sponsored by the American government, the Stanley Trans-Africa Expedition was **financed by two newspapers**, the New York Herald and England's Daily Telegraph. The expedition involved hundreds traveling on foot, 10% of whom were local porters Stanley credits with the success of the expedition. **Mortality rates were high.** Stanley speaks of the difficulty in maintaining order, but was also notorious for cruelty to the indigenous population. The results included conclusive discovery of the source of the Nile through extensive exploration of the Lake Victoria and Lake Tanganyika areas, and exploration of the Congo and Lualaba Rivers.

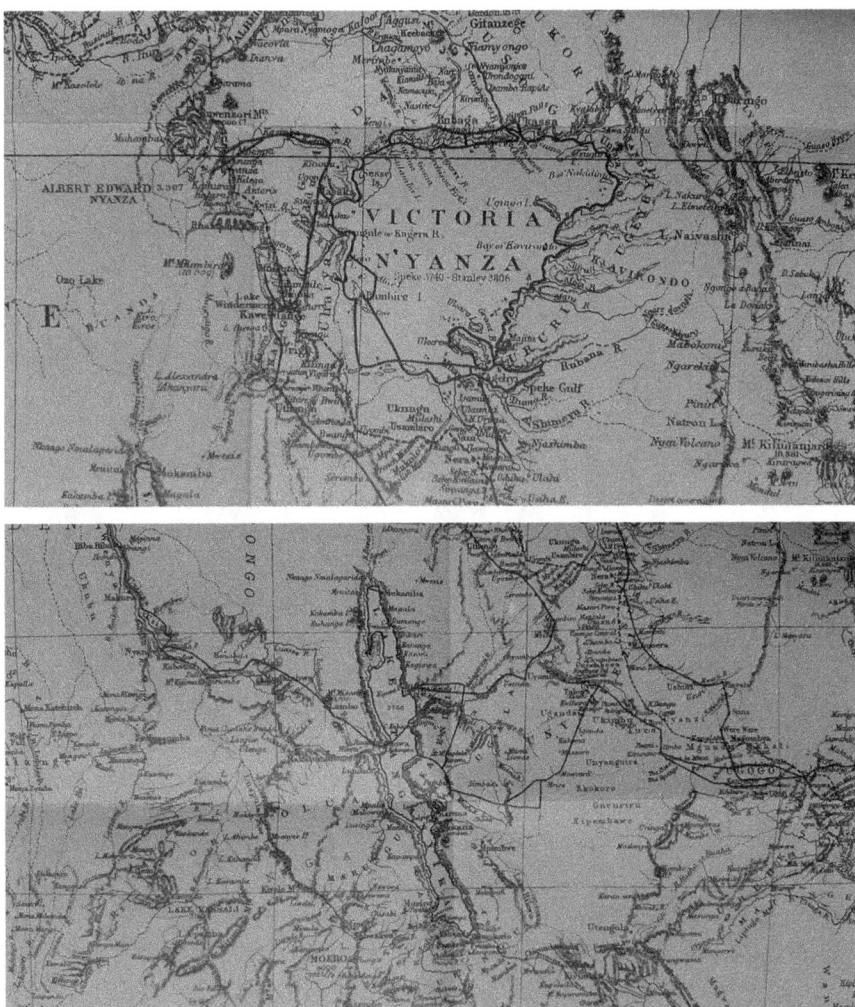

Fig. 2.6 Stanley's routes indicated by heavy black line in Circumnavigation of Lake Victoria found to be the source of the Nile via the White Nile tributary and Lake Albert in the upper left corner (top map) and Exploration of Lake Tanganyika (bottom map). The Lady Alice was used to transport them by water [14]. (Source: Public Domain)

What about transportation and travel? Sectional boats and dugout canoes were typically used for transportation. Many were drowned in the frequent uncharted river rapids, including his last European colleague Frank Pocock. Boats were dismantled, transported around cataracts, and rebuilt as necessary. Travel on water or land here was in fact often deadly. The rain forest was extremely dense with frequent torrential rains. The conditions, including limited availability of toxin- or parasite-free water and food, which included

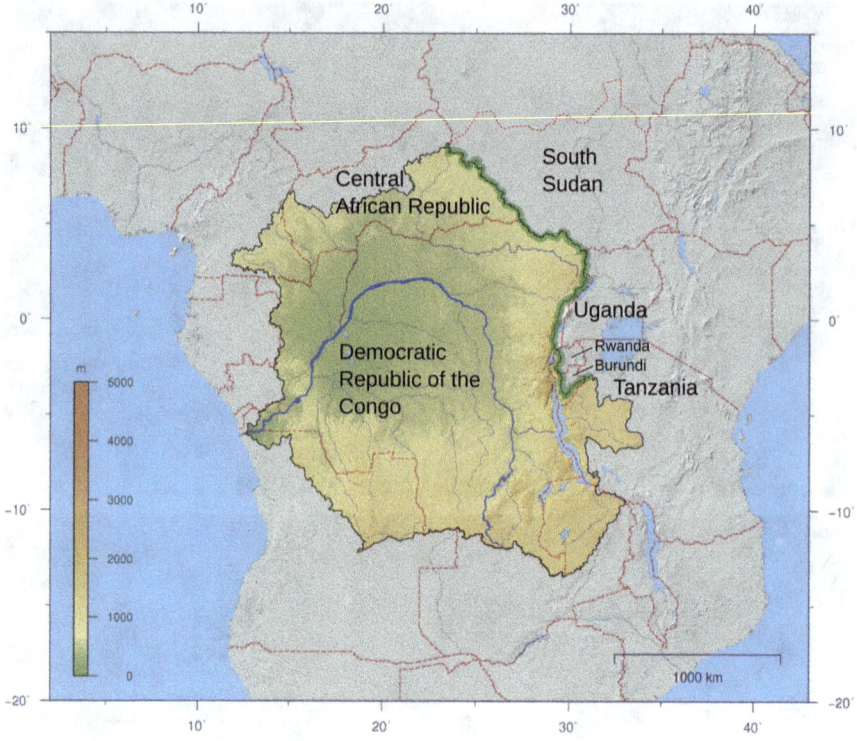

Fig. 2.7 Map of Central African area of Stanley's expedition showing relationship to Nile and Congo Rivers with the divide between Congo and Nile Basins in green [15]. (Source: {annotated} Public Domain)

insects, roots, and fungi, induced hunger, dysentery, and infections, including festering sores. Malaria, and gastrointestinal parasites. Frequent heavy doses of medicines such as quinine also damaged their gastrointestinal tracts. Attacks on the group by indigenous tribes resulted in injury and death by poisoned arrows and spears. Poisonous and frequently deadly animals and plants surrounded them. Many could not go on and had to be left behind. Overall, less than half of those who started with Stanley survived.

Though the character of the participants was different, **the provisioning approach, which included guns and ammunition, portable tools and repair equipment, along with trading supplies, but assumed 'self-provisioning replacement in food and clothing'**, was similar to that used for the Lewis and Clark expedition. Critical supplies included

- navigational instruments,
- materials and equipment to repair clothing and shelter,

- the best available guns and ammunition of the time, including snider rifles, elephant guns, and automatic rifles
- 'sami sami', native trading supplies (cloth, copper wire, beads)
- Tools for jungle hacking and equipment to repair them

The overarching goal was to complete the mapping of central Africa, with particular focus on its lakes and rivers, an exercise already begun by others, as well as to confirm the source of the Nile. The circumnavigation of the two largest lakes, Victoria and Tanganyika, were completed in the course of the expedition. The one tributary from Lake Victoria, previously discovered and considered the likely source of the Nile by John Speke, and further observed by Stanley, was eventually found to drain into the White Nile, a tributary of the Nile, via Lake Albert. To confirm this, Stanley set out to explore the downstream, northern reaches of the Lualaba River, another possible source mapped only in its upper reaches by Livingstone, which they then determined was not connected to the Nile via Lake Albert. Instead, the expedition traced it to the Congo River and the Congo River to the sea 1000 days after the expedition started with half of the original members, and Stanley as the only European survivor. **In terms of meeting the goals of its sponsors, and its contribution to understanding the geography of Africa, the expedition was certainly a success, but the cost, considering over half of those who started the expedition were lost, was high.**

In contrast to the Stanley's Trans-Africa expedition, his subsequent third expedition was sponsored by King Leopold of Belgium, with the goal of claiming the Congo for Belgium, and, because of the political ramifications, was clandestine in nature. Stanley traveled under an assumed name, provisions were shipped via circuitous routes, and communication was via a third party. The approach included building trading stations and trails around the rapids on the Congo River, which Stanley accomplished in 3 years. However, Stanley refused to negotiate treaties that gave absolute sovereignty to the King and thus fell out of favor with the King and many of his European counterparts.

Many artifacts and documents associated with the Stanley and other African expeditions may be found in the Russell E. Train Africana Collection [11–13]. The collection includes 2000 books, photographs; paintings, drawings, and sketches; newspaper articles; manuscripts and letters. Materials from David Livingstone and Theodore Roosevelt, as well as Stanley, are well represented in the collection.

2.4 Scott, Shackleton, and Amundsen South Polar Expeditions, 1901–1916 (Figs. 2.8, 2.9, 2.10, 2.11, 2.12, and 2.13)

Though there were several expeditions to the South Pole launched from 1901 to 1916, known as the Heroic Age of Antarctic Exploration, the Amundsen team was the first to reach the South Pole itself. **All of the teams traveled to**

Fig. 2.8 Top: Shackleton expedition with dog sled on the ice [30]. Source public doman. Bottom: Shackleton expedition with Mongolian ponies on the ice [31]. (Source: Public Domain)

Fig. 2.9 Top: Shackleton's Base Camp where officers and crew shared common space having an evening's entertainment [32]. Source public domain. Bottom: Some of Shackleton's men on the ice sheet [33]. (Source: Public Domain)

Fig. 2.10 Top: Scott's Crew at an Ice Depot [34]. Source public domain. Bottom: Scott in Officers' quarters were kept separate from the crews' quarters [35]. (Source: Public Domain)

Antarctica by boat and then established base camps on the Continent. Three men emerged as 'leaders' of one or more expeditions, including Scott, Shackleton, and Amundsen [16–29].

Scott, a captain in the Royal Navy, who has been described as autocratic, driven, hard-working, and temperamental, was recruited by the Royal Geographic Society to lead the first expedition to Antarctica onboard Discovery in 1901. Shackleton was a member of the expedition. Scott set up

Fig. 2.11 Top: Decorated Interior of Amundsen Base Camp [36]. Source public domain. Bottom: Amundsen Base Camp [37]. (Source: Public Domain)

camp on Ross Island in McMurdo Sound in the winter of 1902. This was the first expedition to explore the surrounding land extensively, using skis and sled dogs. In the summer of 1902, Scott, Shackleton, and Wilson attempted to reach the pole but failed due to sickness and hunger. A relief ship took part of the crew back to England, including Shackleton, but Scott remained for a second winter to perform further geological and zoological studies.

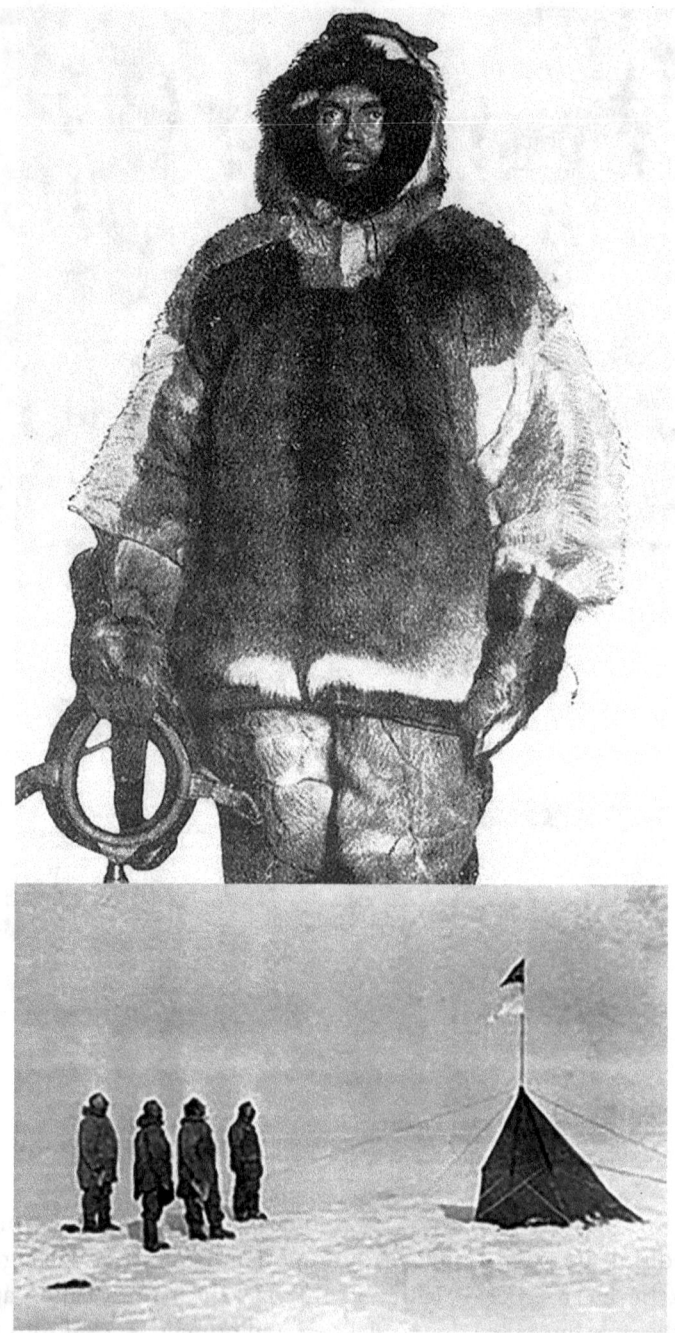

Fig. 2.12 Top: Olaf Blaaland, one of Amundsen's crew, in their arctic gear made of natural and-time-tested material [38]. Bottom: Amundsen camp at South Pole [39]. (Source: Public Domain)

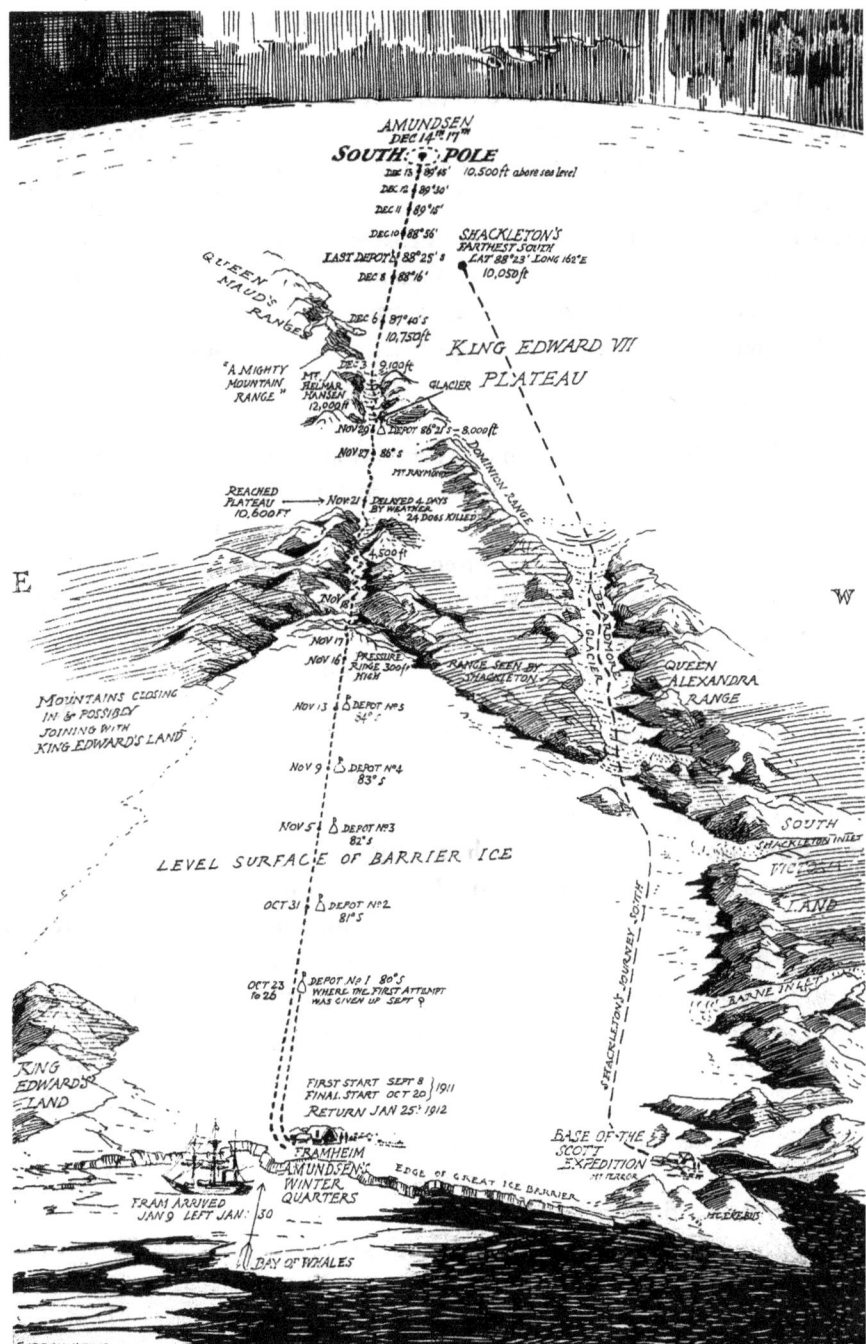

Fig. 2.13 Routes of Amundsen, Shackleton, and Scott [40]. (Source: Public Domain)

Shackleton, although described as extremely stubborn and something of a bully, was known for extraordinary ability to get men to follow him. He had left school at the age of 16 to join the merchant marines and 'seen the world'. His apocryphal sales pitch, in person and appearing as advertisements for a crew, was **"Men wanted for hazardous journey. Small wages, bitter cold, long months of complete darkness, constant danger, safe return doubtful. Honour and recognition in case of success."** Except for his selection of officers Frank Wild and Tom Crean, where he chose Antarctic veterans, **he was not looking for militarily trained or especially skilled men, but for men who struck him as optimistic 'problem-solvers', capable of enhancing the 'esprit de corps'.** That certainly added an element of risk to his expeditions.

Shackleton also led more than one expedition to Antarctica. The very first, the Nimrod Expedition, also known as the British Antarctic Expedition (1907–1909). had the goal of reaching the South Pole first [16–20]. **Shackleton, who had no governmental or institutional support, had to rely on private loans and individual contribution.** Beset by financial problems, Shackleton's preparations were hurried. His ship, Nimrod, was less than half the size of Scott's 1901–1904 expedition ship Discovery, and Shackleton's crew lacked relevant experience. Shackleton's decision to establish his base in McMurdo Sound, close to Scott's old headquarters, in contravention of a promise to Scott that he would not do so, was controversial.

Just as Scott and Amundsen did, Shackleton brought packages of materials from which prefabricated huts could be made, along with tools, guns, and ammunition, and assumed that most of the expedition's protein would be supplied by hunting seal, penguin, and whale. However, to maintain morale over many months of relative 'confinement', reading and musical library and instruments were included as essential supplies.

A couple of months after arrival in the winter of 1908 a Shackleton team, limited in size by the number of available ponies, began the trip to the geographic south pole. The ponies did not survive well on the ice, and eventually man-hauling was used exclusively to move supplies. Although reaching 88° 23′ south and carrying out the first extensive natural history and meteorology studies in Antarctica, the team fell short of the pole by 97.5 nautical miles (181 km). At the same time, an expedition led by Australian Edgeworth David reached the South Magnetic Pole and ascended Antarctica's second highest volcano, Mount Erebus. The Nimrod Expedition was by far the longest southern polar journey and achieved the closest convergence to the pole at that date. Despite the lower profile of Shackleton's expedition, its achievements made Shackleton a national hero. The scientific team carried out extensive geological, zoological, and meteorological work. Shackleton's transport arrangements, based on Manchurian ponies, motor traction, and sled dogs, were innovations which, despite limited success, were later copied by Scott in his ill-fated Terra Nova [29] expedition.

Shackleton launched a second expedition (1914–1916), the British Imperial trans-Antarctic Expedition, with the goal crossing the continent, from a base on the Weddell Sea to McMurdo Sound via the south pole. However their boat, the Endurance, was trapped in the ice. The crew drifted on ice floes and ultimately escaped in boats to Elephant Island in the South Shetland Islands. Shackleton and five other crew members used a whale boat to cross 800 miles of open ocean to South Georgia, where there was a whaling station, to seek help. They then had to cross the island overland because the station was on the other side. Shackleton then then led expeditions to rescue his remaining crew on Elephant Island. The fact that not one of his crew members died throughout this ordeal is a major testimony to his leadership. Shackleton died of a heart attack on South Georgia Island as he was starting another expedition with the goal of circumnavigating Antarctica in 1921.

The Amundsen and Scott Antarctic Expeditions (1910–1913) were focused on the same goal of reaching the South Pole first, though Scott had a broader goal of exploring polar regions and Amundsen had initially planned a north polar expedition via arctic drift in an icebound ship before Peary and Cook announced their arrival there in 1909 [22–29]. Amundsen, described as thorough, meticulous, and orderly, had extensive experience both as a seaman and as a cross-country skier. Both Scott and Amundsen had been on earlier Antarctic expeditions. As first mate, Amundsen had experienced winter entrapment in a ship on the ice around Antarctica in a much earlier (Belgian) expedition and took notes on the impact on the crew (starvation, scurvy, insanity, depression) as well as how to avoid this situation in future expeditions. Following this, Amundsen was the first to find a route through the northwest passage.

Both the Amundsen and Scott Antarctic expeditions, though privately financed to a large extent like Shackleton's, also had support from their respective governments, and Scott utilized newspaper advertisements to obtain corporate sponsors in return for purchasing and announcing the use of their brand names for the expedition.

Amundsen and Scott expedition teams were led by naval officers and selected for crucial survival skills, including navigation, carpentry, handling the selected transportation equipment (sleds, skies, dogs), medicine, engineering, and, for Amundsen, an experienced ship's cook, an approach similar to that used by the Lewis and Clark expedition.

Amundsen specifically looked for 'resourcefulness' in the crew selection process. Their leadership styles were different: more flexible Amundsen relying on advice from others with more experience and being more cautious when dangers arose, and less flexible Scott relying more on courage and luck. Amundsen arrived with a party of 19 men, with 4 taken to the pole, whereas Scott arrived with 16 men, with 5 taken to the pole.

Scott arrived first, establishing a base camp at McMurdo Sound at Cape Evans (15 miles north of Hut Point, Scott's 1902 base), with the means to travel from the base camp by motorized sledge, ponies, and human or dog hauled sled, similarly to Shackleton. This was known as the Terra Nova Expedition. Scott's group constructed the prefabricated hut rapidly and began the depot-laying operation in the winter of 1911–1912. From Safety Camp at the edge of the ice barrier, the goal was to establish a number of depots, each closer to the pole, with the last one being One Ton depot within ten degrees of the pole. Meanwhile a team led by Campbell, another member of Scott's expedition, explored Victoria Land to the east and encountered the Amundsen expedition at the Bay of Whales. Their offer to help was declined and the team returned to report the news to a very unhappy Scott. After a series of mishaps leading to the death of many of the ponies, Scott returned to Cape Evans for wintering over. Unlike Shackleton's camp, the base camp was divided to keep officers and crew in separate quarters. During 1911–1912, several successful geological and zoological expeditions were launched to the west, east, and north. In September 1911, Scott launched the poleward expedition, planning to ascend the glacier and cross the polar plateau. The motorized sledges failed. Manhauling became the normal cargo hauling mode. On January 17 1912 with a final team of 5, Scott reached the pole to discover Amundsen had already been there. They headed back, making it to the summit of the glacier in 3 weeks, where Evans died. They never made contact with the dog teams on the way back as planned. They found inadequate fuel for warming at the depots, and were slowed down by serious frostbite. All members of the polar party perished, including Scott. **The failures have been blamed on Scott's authoritarian leadership style, his poor judgement of character, the far worse than normal persistent weather conditions resulting in frequent blizzards, and the disregard of Scott's orders to bring the dog teams back for use of the polar party on their return.**

Amundsen set up based camp on the Ross Ice Sheet in the Bay of Whales, 60 miles closer to the pole, and planned to travel by ski and dog sled. Both Scott and Amundsen teams were heavily provisioned, though Amundsen planned on supplementation with seal meat, and created provision depots along the planned route in advance.

The nature of Amundsen's ship used to travel to and from the Antarctic contributed greatly to the expedition's success. The ship, Fram, was designed and built in 1891–1893 by Norway's leading shipbuilder and naval architect, Colin Archer, to exacting specifications: it needed to withstand prolonged exposure to arctic conditions, and to be able to endure and slip through arctic ice it had extra crossbeams and braces made of the hardest timber known (greenheart) and a rounded hull. Although its design caused it to wallow and roll uncomfortably in the open ocean, Fram proved most successful in dealing with the ice pack.

After setting up base camp, Amundsen had the ship return to Buenos Aires for reprovisioning and then engage in an extensive oceanographic cruise in Antarctic waters before returning to base camp. Meanwhile, those on the ice created the provision depots before making the first attempt for the south pole after the ship returned. Amundsen learned much about the performance of the equipment and men during the provisioning journeys, finding travel much like that on conventional glaciers. The dogs had difficulty in pulling particularly heavy sleds, so Amundsen took more dogs and fewer men on the expedition to the South Pole. The men spent polar winter before the first attempt redesigning and lightening the sleds and modifying their personal gear for improved performance. Sources of Vitamin C (to prevent scurvy, which did occur on both Shackleton and Scott expeditions) included bottled berries and seal meat. Vitamin B in whole meal bread made with fresh yeast was added to their diets. Amundsen was worried about competing with the apparent capability of Scott's motorized sledges, and thus, despite the extreme cold, made the first attempt to reach the South Pole which met with near disaster, including frostbitten dog paws and human extremities in the Spring of 1911. Amundsen dug in until October 2011, and, on the second attempt, Amundsen's team, making 15 nautical miles a day, reached the pole (1 month before Scott) in December of 2011, and returned in 100 days (early 1912). They discovered and traveled via the Axel Heiberg glacier which led them to the central polar plateau and the pole itself. The team butchered and ate half of their dogs at the summit of the glacier before making the final push to the pole. Once on the plateau, going was made difficult with the presence of many crevasses. After ascending for three days to establish their geographic position, they returned, traveling during polar 'night' (even though they had daylight all the time) to keep the sun at their backs and prevent snow blindness. Ski travel was easier than sled travel on the descent, when it was harder for sleds to stop in time to avoid the crevasses. **The team expertise with skis and dog sleds played a major role in their success.** Scott's team reached the pole 1 month later.

Amundsen's equipment fully utilized features that were proven to work in arctic environments, minimized resources needed for assembly and maximized nourishment. The ski boots were the result of 2 years of testing, and the polar suits included suits with traditional materials, such as northern Greenland sealskin, reindeer skins, wolf skin, Burberry cloth and gabardine, in styles of the Netsilik Inuit. The extra-long skies and sledge runners were made from hickory. The tents had built in floors and a single pole and the stoves were compact Swedish Primus. Amundsen developed a nourishing pemmican, insisted on regular supplementation with seal meat, supplied many books, musical instruments, records and a gramophone, as well as plenty of wine and spirits.

2.5 Comparison to Apollo Lunar Expeditions

The Moon is a place even more remote and dangerous than Antarctica. At the beginning of the 'age of space exploration' the culture-wide assumption was that expeditions there would be exceedingly dangerous and involve loss of life. That scenario was clearly assumed in the 1959–1960 American TV series, Men into Space. More recently, Neil Armstrong stated in a 2012 interview with Alex Malley, Chief Executive of CPA Australia: "A month before the launch of Apollo 11, we decided we were confident enough [to] try an attempt on a descent to the surface. I thought we had a 90% chance of getting back safely to Earth on that flight but only a 50-50 chance of making a landing on that first attempt." This was a primary reason why the commitment by Kennedy for a human landing on the moon in less than a decade was so daunting. At the same time, as in the heroic age of exploration, the assumption was that the sacrifice would be necessary and was worth the price. Then as now, a strong political component, a race for prestige, power and sovereignty, was a component. In what other ways was the Apollo program, the earliest program of lunar expedition culminating in human exploration of the surface, similar to or different from earlier explorations? And how did it influence all subsequent exploration?

There are obvious differences between the Apollo missions and earlier expeditions. (For context, see references in Chap. 1). One striking difference with earlier high impact expeditions was the number of people involved. In terms of total numbers of individuals and organizations involved in providing all manner of support, the Apollo program was larger by orders of magnitude: hundreds of thousands all over the world as opposed to hundreds for earlier expeditions. On the other hand, in terms of the number actually on site,

Apollo had three as opposed to tens or hundreds, though for the polar expeditions, the team that actually went to the pole was a very small group, certainly less than ten. Another difference was contact with the outside world. Though at orders of magnitude greater distance, the Apollo astronauts were often in communication with mission control and heard by the entire world. Generally speaking, earlier expeditions were incommunicado and had contact with the outside world only by prearrangement. Finally, all of the earlier expeditions relied on and were successful based on the use of traditional practices or use of materials. That included deliberate embracing of 'living off the land' practices of indigenous (or in the case of the polar expeditions, Arctic) people. The Apollo program relied on the development of at least ten entirely new technologies rapidly developed from relatively loosely defined concepts, including orbital flight; Earth gravity escape; orbital communication, rendezvous and docking, navigation and tracking; deep space communication, navigation and tracking; lunar flyby, rendezvous, orbit, and landing; and life support.

The Apollo program (1961–1972) (Table 2.1) and precursor development programs occurring throughout the sixties that made it possible (Ranger, Lunar Surveyor, Lunar Orbiter, Mercury, Gemini) were financed by the American government, NASA, through a new kind of partnership with industry and academia. Considering the great distances, isolation, and extraordinary challenges in traveling to and surviving on the target, and the long lead time for return on capital investment, government sponsorship was the only viable model at that time. The technologies needed to land a human being on the Moon had not been developed when the commitment was made, as mentioned above. Engaging in activity on the lunar surface would require a portable life support system, which turned out to be an uncomfortable pressure suit with minimal manual dexterity, allowing travel on foot or via an unpressurized vehicle. Both jet packs and wheeled rovers were under consideration initially, but the wheeled rover was considered less risky and prevailed. Travel in an unpressurized vehicle in a low gravity, high temperature, high radiation vacuum had never been done before.

Table 2.1 Apollo program and its precursors

Program	Dates	Notes
Apollo	1960–1972	10 out of 11, first crewed landing
Mercury	1961–1964	6 out of 6, first us crewed program
Gemini	1961–1966	10 out of 10, rendezvous, docking, EVAs
Ranger	1964–1965	Hard landers, 3 out of 9
Surveyor	1966–1968	Soft Landers, 5 out of 7
Lunar Orbiter	1966–1967	Orbiters, 5 out of 5

Nevertheless, a 'can do' attitude prevailed, as it had for the other expeditions. Analogously as well, leaders were naval, or air force, officers with experience as test pilots, and candidates were hand-picked through the highly competitive astronaut selection process for their technical expertise as well as ability for problem-solving and endurance under very adverse circumstances. Unlike earlier expeditions, the Apollo program involved small crews of astronauts, who had received frequent simulation training, and were supported by extensive personnel in mission ground control, were sequentially launched to selected targets on the lunar surface. But like earlier expeditions, particularly those of the twentieth century, media coverage was considered crucial, to assure public support for these publicly financed campaigns.

The differences in expedition costs reflect these differences as well. The Lewis and Clark Expedition, based on actual calculation of expenses, cost $40,000 [41], about $50,000,000 in today's dollars, considerably less than the $25,000,000,000 {$250,000,000,000 in today's dollars) [42] invested in the Apollo Program, although it should be kept in mind that the Lewis and Clark team 'lived off the land'. The London Daily Telegraph and New York Herald officially contributed £12,000 in about 1870 [43], amounting to about $12,000,000 in current dollars, to Stanley's Trans-Africa Expedition, likely a great underestimate of total cost, as that amount is not based on actual expenses and Stanley's team also 'lived off the land'. Finally, Shackleton's costs were estimated to be about £100,000 [44] in 1910, or about $65,000,000 in today's dollars. The five expeditions to the Antarctic (Amundsen, two for Scott, two for Shackleton) combined costs would have been approximately $325,000,000. Dollar calculations are based on conversions from pounds sterling to dollars and the worth of the dollar or pound in the year indicated [45, 46].

The lunar surface, 400,000 km away across deep space and unprotected by an atmosphere, consists of regolith generated from meteoritic bombardment of its ancient crust, and 'weathered' from interactions with charged particles and micrometeorites. Thus, the lunar surface is dominated by impact craters, and volcanic plains which have filled the larger impact basins. Both impact craters and volcanic terranes reveal the nature of underlying stratigraphy, according to the principle of superposition. The most recent eruptions of volcanic materials (magma erupting more gently as lava or explosively as pyroclastic deposits) are exposed closest to the surface, overlying older deposits. Craters reveal underlying strata in reverse, with the deepest strata of underlying rock exposed closest to rim and shallowest further from rim, otherwise known as reverse stratigraphy. In both cases, stratigraphic history is revealed when geologists perform traverses, sampling and obtaining various measurements across surface features. Thus, from the beginning, the focus has been on sampling rock and regolith from the lunar surface and monitoring the surrounding environment [47]. Although the likelihood of discovering life on the Moon was considered extremely small, protocols were used to determine biogenicity in collected samples. Thus, planning for sampling along geological traverses on the lunar surface became an important part of the preparation for the Apollo missions to the Moon [48–52]. This consideration also drove the need for precursors to establish the scale and extent of variation in the regolith and its relationship to underlying rock on a scale as small as a meter or less at landing sites. Understanding at that scale was necessary in order to plan, perform simulations, and train astronauts for science operations [53–55]. The design, deployment and methodology for use (sampling strategy and documentation) of geological tools and instrument packages were developed and tested on the basis of their adequacy to characterize site geology in the broadest context [48, 55].

An originally unforeseen consequence of the lunar surface bombardment process was that it generated regolith with very unEarthlike and challenging handling properties (Fig. 2.14) [56]. A large portion of lunar soil consists of glassy elongated shards with highly complex surface structure behaving like 'abrasive velcro'. Everything that came in moving contact with it,

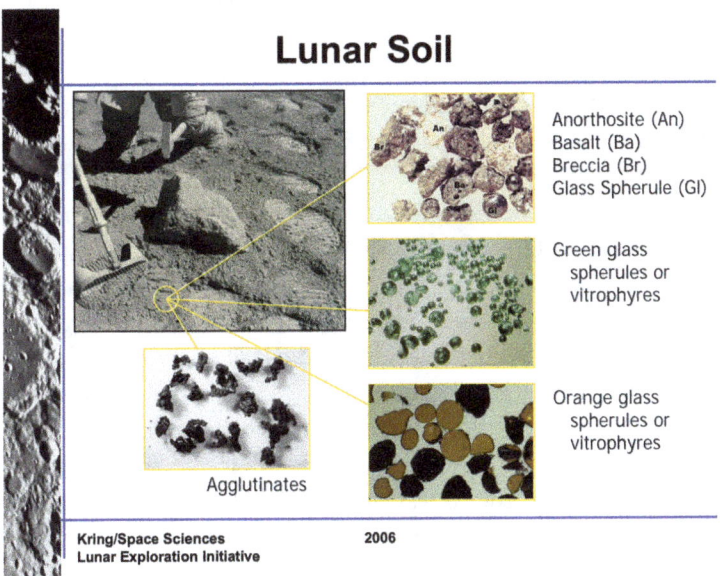

Fig. 2.14 Distinctive shape, size distribution, and components of lunar soil as discussed in the text [56]. (Credit: NASA Lunar Initiative Briefings)

including the pressure suits and rovers, were covered with it. Applying surface pressure to the regolith, compressing the grains, caused them to align, and increased the apparent hardness of the regolith. Attempts to brush away dust caused abrasion of pressure suits which would ultimately have caused puncturing. The astronauts managed to kludge dust covers for the rover wheels from their manuals. How this affected the tool developed for sample collection, and how difficulties were overcome, will be discussed in subsequent chapters.

The inclusion of a team of geologists 'in the backroom' during astronaut lunar surface EVAs, acting in an advisory capacity, was an important addition to the architecture. This 'foot in the door' approach encouraged openness and rapid dissemination of information to the greater community as well [57]. That resource did not mitigate the need for extremely capable crew members with exquisitely good training in field geology, resourcefulness, extraordinary problem-solving skills, and the ability to assess and articulate the surrounding context quickly, to describe it with the appropriate level of detail, and to collect sufficiently representative samples that when returned to Earth would considerably advance the understanding of the Moon by the scientific community.

How could science requirements be met while dealing with severe, necessary constraints? That question drove the development of the unique Apollo Field Work Style, which will be addressed in subsequent chapters. The salient features include the following:

- Despite the fact that this had never been done before, robotic precursors providing 'on the ground' images and orbital coverage at up to 1 m resolution for potential landing sites would be essential.
- Although pictures 'on the ground' would be available from the Lunar Surveyor lander and later from earlier Apollo missions, and equipment was designed when lunar soil properties were unknown, the astronauts would have to rely on well-planned training and simulations for experience of what field work at a particular site would entail,
- No return visits or 'do overs' would be possible, so the astronauts had to be successful at capturing the geological character site visited accurately and succinctly the first and only time, despite the fact that they had very little time (3 EVAs amounting to less than 24 h in the field for the later J missions) and very little onboard mass (100 kg) available for returned samples.
- The astronauts were restricted to tens of kilometers in the rover and normally tens of meters on foot, as well as the limited dexterity spacesuit wear afforded.

These constraints affected every aspect of the approach to surface activities well as field hardware design, and resulted in approaches far different from conventional geological field work.

References

1. Jones, Landon, 2002, The Essential Lewis and Clark, Ecco, 224 pages.
2. Preparing for the Lewis and Clark Expedition, Jefferson Monticello. https://www.monticello.org/thomas-jefferson/louisiana-lewis-clark/preparing-for-the-expedition/
3. Moulton, Gary, The Lewis and Clark Expedition Day by Day, 2018, University of Nebraska Press, 774 pages.
4. Fritz, Harry, Discover Lewis & Clark: Synopsis of the Expedition, https://Lewis-Clark.org
5. National Park Service, Ink and the Lewis and Clark Expedition, 2019. https://www.nps.gov/articles/ink-and-the-lewis-and-clark-expedition.htm visited 010125.
6. National Park Service, William Clark: A Master Cartographer, 2023. https://www.nps.gov/articles/william_clark_cartographer.htm. visited 010125.
7. Paxson, Detail of painting Lewis, Clark, and Sacajawea at Three Forks, Public Art in Lobby of Montana House of Representatives, 1912. https://mhs.mt.gov/education/Capitol/Art/House-Lobby
8. Smithsonian National Museum of American History, Blunderbuss ca 1809-1810, Behring Center, Washington DC. https://www.loc.gov/exhibits/lewisandclark/lewis-landc.html visited 010125.
9. Clark, William, Track across western North America from Mississippi to Pacific by executive order of US government, 5&6. Bradford and Inskeep. Library of Congress Geography and Maps Division, 1814. https://www.nps.gov/articles/william_clark_cartographer.htm visited 010125.
10. Montana Historical Society, Great Falls, 1880. https://www.nps.gov/articles/the-great-falls-according-to-lewis.htm visited 010125.
11. The First Trans-Africa Expedition, https://en.wikipedia.org/wiki/Henry_Morton_Stanley
12. Henry Morton Stanleys Unbreakable Will, https://smithsonianmag.com/history/henry-morotn-stanleys-unreakable-will-99405/
13. Henry Morton Stanley, 1879, Through the Dark Continent, two volumes, New York, Harper Brothers, 657 pages.
14. Henry Morton Stanley's First trans-Africa Expedition. https://www.wikiwand.com/en/articles/Henry_Morton_Stanley%27s_first_trans-Africa_expedition. visited 010125.

15. Wikimedia commons, Imagico and Aymatth2, Congo Basin, Congo-Nile Divide. https://commons.wikimedia.org/wiki/Category:Congo_Basin visited 010125.
16. The Nimrod Expedition. https://en.wikipedia.org/wiki/Nim_rodExpedition
17. Ernest Shackleton. https://en.wikipedia.org/wiki/Ernest_Shackleton
18. Fiennes, Ranulph, 2022, Shackleton, Pegasus Books, 413 pages.
19. Larson, Edward, 2011, An empire of Ice: Scott, Shackleton, and the Heroic Age of Antarctic Science, Yale University Press, 360 pages
20. Hunt, Rebecca, 2014, What do the diaries of Shackleton and Scott Reveal, The Guardian, March 22, 2014, UK. https://www.theguardian.com/books/2014/mar/22/diaries-ernest-shackleton-captain-scott-reveal
21. Terra Nova Expedition. https://en.wikipedia.org/wiki/Terra_Nova_Expedition
22. History of Scott's Expedition, Antarctic Heritage Trust. https://nzaht.org/conserve/explorer-bases/scotts-hot-cape-evans/history-of-scotts-expedition/
23. Robert Falcon Scott. https://en.wikipedia.org/wiki/Robert_falcon_Scott
24. South Pole Exploration: Robert Falcon Scott, 1901-1904, Royal Museums Greenwhich. https://www.rmg.co.uk/stories/topics/south-pole-exploration-robert-falcon-scott-1901-04#:~:text=Scott's%20first%20expedition%2C%201901%E2%80%9304&text=Sir%20Clements%20Markham%2C%20the%20influential,Discovery%20on%2031%20July%201901
25. Ponting, Herbert, 2024, The Great white South, or with Scott in the Antarctic, Daredevil Books, 452 pages.
26. Amundsen's South Pole Expedition. https://en.wikipedia.org/wiki/Amundsen%27s_South_Pole_expedition
27. The Race to the South Pole: Scott and Amundsen, Royal Museums Greenwich. https://rmg.co.uk/stories/topics/race-south-pole-scott-amundsen
28. Huntford, Roland, 2010, Race for the South Pole: The Expedition Diaries of Scott and Amundsen, Continuum, 353 pages.
29. Amundsen, Roald, 1911, "The History of the South Pole". The South Pole, An Account of the Norwegian Antarctic Expedition in the "Fram," 1910–1912. **1**. London, United Kingdom: John Murray. Retrieved 10/10/24.
30. Hurley, J.F., A seven-dog sledge team on the ice with unidentified Endurance crew member (4793354178), 1915. https://picryl.com/media/a-seven-dog-sledge-team-on-the-ice-with-unidentified-endurance-crew-member-5cc6e1. visited 010125.
31. Shackleton, E.H., Shackleton Nimrod 28, 1908. https://commons.wikimedia.org/wiki/File:Shackleton_nimrod_28.jpg. visited 010125.
32. Shackleton, E.H., Shackleton nimrod 32, A group of people sitting around a table, 1907-1909. https://picryl.com/media/shackleton-nimrod-32-62a322. visited 010125.
33. Shackleton, E.H., Ernest Shackleton on his South Polar Expedition, The Public Domain Review, 1910. https://publicdomainreview.org/collection/ernest-shackleton-on-his-south-polar-expedition-1910/. visited 010125.

34. Ponting, H., Depot laying party, Scott Expedition, 1911. https://commons.wikimedia.org/wiki/File:Depot_laying_party_1911.jpg. visited 010125.
35. Ponting, H., Captain Robert F. Scott writing in his diary, Scott Expedition, 1910. https://loc.getarchive.net/media/captain-robert-f-scott-writing-in-his-diary
36. Amundsen, R.S., Saloon decorated for Christmas eve., 1912. https://no.wikisource.org/wiki/Fil:Amundsen,Roald-Sydpolen_I-1912-p283.jpg visited 010125
37. Amundsen, R.S., Framheim, 1911. https://en.wikipedia.org/wiki/file?framheim,_february_1911.jpg. visited 010125.
38. Amundsen R.S., Bjaaland in Winter Clothing, Amundsen South Pole Expedition, 1911. https://en.wikipedia.org/wiki/File:BjaalandWinterclothing1911.jpg. visited 010125.
39. Bjaaland, O., Members of Roald Amundsen South Pole Expedition 1910-12 at the pole itself, 1911. https://en.wikipedia.org/wiki/Amundsen%27s_South_Pole_expedition#/media/File:At_the_South_Pole,_December_1911.jpg. visited 010125.
40. Home, G., Map of Roald Amundsen's South Pole Expedition, in Murray, J., The History of the South Pole, The South Pole, An account of the Norwegian Antarctic Expedition in the Fram, 1910, 1912, 1, 1911. https://en.wikipedia.org/wiki/Amundsen%27s_South_Pole_expedition#/media/File:Gordon_Home's_Map_of_Amundsen's_South_Pole_Expedition.jpg. visited 010125.
41. https://lewis-clark.org/the-trail/eastern-beginnings/estimate-of-expenses/
42. https://www.planetary.org/space-policy/cost-of-apollo
43. https://www.telegraph.co.uk/travel/destinations/africa/congo/articles/Henry-Morton-Stanley-in-the-Congo/
44. https://en.wikipedia.org/wiki/Imperial_Trans-Antarctic_Expedition
45. https://www.measuringworth.com/datasets/exchangepound/
46. https://www.officialdata.org/us/inflation/1803?amount=1
47. Beatty, D., Taking Science to the Moon, Johns Hopkins University Press, Baltimore, 2001.
48. Benjamin, P., Revision of Operational Constraints for J Mission Traverse Planning, Case 320, Bellcom 2032-PB-mp, 1970.
49. Kain, R., R. Koppa, J. Olmsted, T. Montgomery, Preliminary Apollo 16 Preliminary Lunar Surface Procedures, NASA TMX70010, 1971.
50. Bland, D., R. Blevins, J. Olmsted, R. Koppa. Final Lunar Surface Procedures, Volumes 1 and 2, 1972.
51. Wood, B., Apollo 17 Traverse Planning Data, 3rd Edition, 1972, reissued in 2008.
52. Wood, B. Apollo Program Summary Report, JSC–09423, http://history.nasa.gov/alsj/APSR-JSC-09423.pdf, 1975.
53. Schaber, G., Chronology of Activities from Conception through the End of Project Apollo 1960-1973, USGS Astrogeology Branch, Open File Report 2005-1190, http://pubs.usgs.gov/of/2005/1190/, 2005.
54. Phinney, William, Science Training History of the Apollo Astronauts, NASA/SP-2015-626, 328 p., 2012.

55. Ables, P., Time and motions required to perform an active seismic experiment proposed for the first Apollo landing, USGS Technical Letter, Astrogeology 12, 1967.
56. Kring, D. Parameters of Lunar Soils, Lunar Exploration Initiative, 2006. https://www.lpi.usra.edu/science/kring/lunar_exploration/briefings/lunar_soil_physical_properties.pdf
57. Hammond, A., Lunar Science: Analyzing the Apollo Legacy, Science, 179, 4080, 1313–1315, 1973.

3

The Context: Robotic Precursors Ranger, Surveyor, Orbiter

3.1 Robotic Precursors

While the Mercury, Gemini, and Apollo programs developed core technologies, selected and equipped astronauts for deep space exploration, other robotic precursor programs provided the reconnaissance data critical for planning, simulating, and executing Apollo field activities (e.g., [1–4]) (Table 3.1). The earliest, Ranger (1961–1965), implemented by NASA JPL, established the basic capabilities for navigation, tracking, and communication from the Earth to the Moon. Lunar Orbiter (1966–1967), implemented by NASA Langley, confirmed the capability for orbital insertion and provided pictures of the potential landing sites at nominal resolution of 1 meter. Lunar Surveyor [1966–1968], implemented by NASA JPL, verified that soft landing on the moon was possible, established that the regolith could support a lander and its crew, and provided the first pictures and compositional data of the lunar surface in situ. Aerial reconnaissance data with comparable resolution were obtained for selected terrestrial sites where simulation and astronaut training would be performed.

Table 3.1 Apollo program and its precursors

Program	Dates	Notes
Apollo	1960–1972	10 out of 11, first crewed landing
Mercury	1961–1964	6 out of 6, first us crewed program
Gemini	1961–1966	10 out of 10, rendezvous, docking, EVAs
Ranger	1961–1965	Hard landers, 3 out of 9
Lunar Orbiter	1966–1967	Orbiters, 5 out of 5
Lunar Surveyor	1966–1968	Soft Landers, 5 out of 7

3.2 The Ranger Program

The first precursors were the Ranger spacecraft (Figs. 3.1, 3.2, 3.3, and 3.4). The Ranger project [5–7], had the particular challenge of developing the capability for deep space operation, including remote navigation, tracking, communication, and spacecraft component automation, in a working environment where the feedback and priorities were constantly changing. JPL was chosen to lead the Ranger Program, based on Pickering's pioneering work in radio telemetry and inertial guidance utilized in the Explorer and Pioneer satellite programs in 1958 and 1959. JPL had provided a TV camera with a photoelectric trigger. The discovery of the Van Allen Belts increased interest in

Fig. 3.1 Ranger Two at JPL. (Credit: NASA/JPL-Caltech)

Mission Details

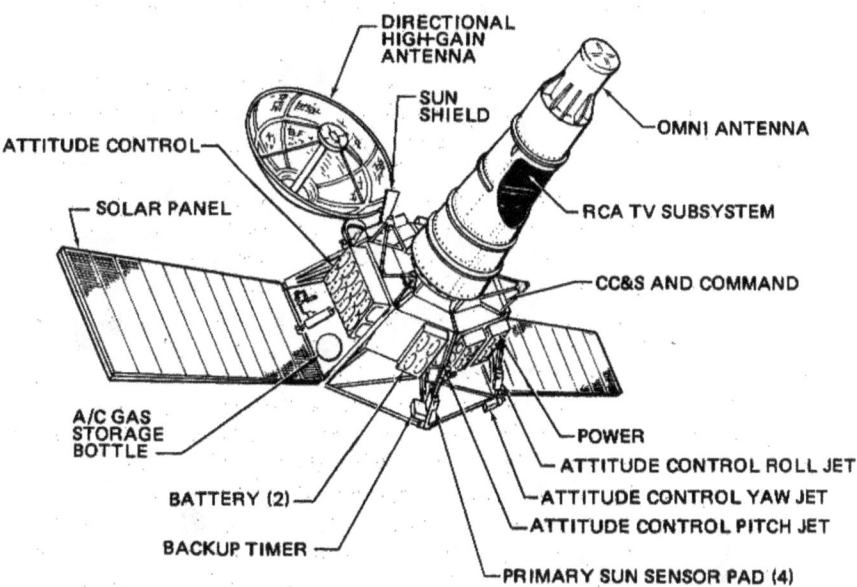

Key Dates
Launch: July 28, 1964 / 16:50:07 UT
Lunar Impact: July 31, 1964
Nation: United States of America (USA)
Objectives: Lunar Impact
Spacecraft: P-54 / Ranger-B Spacecraft
Mass: 806 pounds (365.6 kilograms)
Mission Design and Management: NASA / JPL
Launch Vehicle: Atlas Agena B (Atlas Agena no. 9 / Atlas D no. 250 / Agena B no. 6009)
Launch Site: Cape Canaveral Air Force Station, Fla. / Launch Complex 12
Scientific Instruments
1. Imaging System (Six TV Cameras)

Fig. 3.2 Ranger 7 Mission Details. (Credit: NASA/JPL-Caltech)

in situ measurements for radiation, fields and particles ('sky' science); thus, subsequent satellites were equipped with cosmic ray and micrometeorite detectors. 'Sky' science instruments, which provided useful measurements in suborbital and balloon tests prior to orbital rocket launches, dominated early payloads. Surface science instruments, including telescopes and radar, required further maturation to acquire higher resolution images and compositional

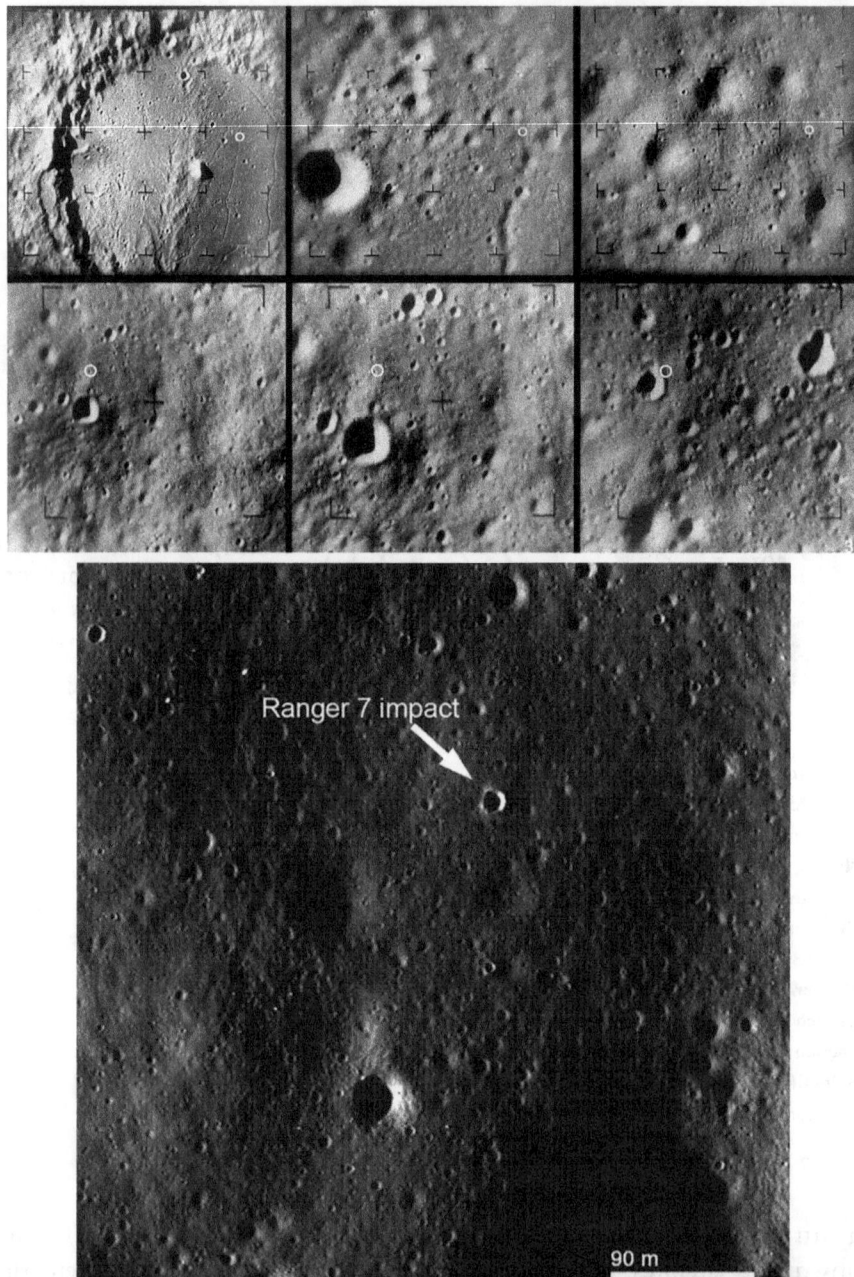

Fig. 3.3 Top: Ranger 9 Approaching the Moon (Top left to right, then bottom left to right. (Credit: NASA/JPL-Caltech). Bottom: First successful Ranger lander (7) landing site. (Courtesy of NASA)

3 The Context: Robotic Precursors Ranger, Surveyor, Orbiter

Fig. 3.4 JPL Mission Control for Ranger Program 1964. (Credit: NASA/JPL-Caltech)

data from remote objects, as well as advancement in capabilities for remote rendezvous, automated orbit, and soft landing. From the beginning, JPL pushed the envelope in support of these developments. Newell, Jastrow, and Urey were responsible for shifting the focus to include the Moon, thus providing a rationale for the Ranger program, as well as a basis for including non-visual compositional measurement instruments (X-ray, Gamma-ray) in post-Ranger lunar payloads.

The first spacecraft bus concept was developed as part of the Ranger Spacecraft. For optimal long-distance control and communication, Ranger was 3-axis stabilized, unlike many earlier spacecraft, which were spin-stabilized. NASA utilized university 'experts' as principal investigators for state-of-the-art instruments designed to measure radiation (gamma-ray lines), plasmas, energetic particles (indeed, sky science). There was interest in magnetometers, to measure magnetic fields, as well as seismometers, related to Moon gravity and mass distribution. The TV camera was to be activated within 4000 km of the lunar surface to achieve 3 m spatial resolution, a great improvement over anything achievable from Earth, and a radar altimeter was to be used initiate payload separation. More onboard automated remote control (avionics), redundancy, mass, and power were required to deal with a remote target as well.

The Deep Space Network, one of NASA's greatest and most ingenious assets, was itself developed to support Ranger and subsequent lunar program flight operations. The network of globally distributed ground stations with massive, steerable, sensitive antennas would always be able to command, determine position (navigate and track), and receive telemetry from deep spacecraft. Using predetermined ephemeris to locate a spacecraft at any given time, they would point and lock on to the spacecraft, and then use the radio signal from the spacecraft itself to track and simultaneously transmit and receive signal from the spacecraft.

The Ranger Spacecraft were built in three blocks (Table 3.2): Block one consisted of two lunar flybys. Due to failures of the Atlas Agena launch system, neither of these were successful. Yuri Gagarin was successfully launched into space before Block two was flown, further shifting emphasis on science that would enable human landing on the moon. Block two were to be three hard landers each equipped with radiation detector (surface composition), capsule containing a seismometer (surface stability), radar altimeter (ranging as well as surface reflectivity), and cameras to take pictures at different times on the way to the lunar surface. These cameras were designed to provide close-ups of the surface (particularly to support design of human crewed vehicle landing gear) by improving spatial resolution by more than three orders of magnitude. None of these were successful. Block 3 (Fig. 3.5), Rangers 6–9, were equipped more redundantly for sequential operation as Apollo landing dates approached. Rangers 7–9 succeeded. Each had TV cameras to provide up to 0.5 m resolution, to produce video data to be recorded on 35 mm film and magnetic tape at Goldstone. Still scientists were not entirely satisfied at this point, due to the significant program delays, as well as the refusal of program management to approve additional payload to

Table 3.2 Ranger missions

#	Launch	Goal	Issues
1	8/23/61	Block 1: Test prototype in Earth Orbit	Could not stabilize
2	11/18/61	Block 1: Test prototype in Earth Orbit	Could not stabilize
3	1/26/62	Block 2: Lunar probe/impactor	Missed Moon
4	4/23/62	Block 2: Lunar probe/impactor	Failed when hit Moon
5	10/18/62	Block 2: Lunar probe/impactor	Missed Moon
6	01/30/64	Block 3: Lunar probe/impactor/imager	Cameras disabled, redesigned
7	07/28/64	Block 3: Copernicus, Mare Cognitum impact	None
8	02/17/65	Block 3: Oceanus Procellarum, Mare Nubium, Mare Tranquillitatis impact	None
9	03/21/65	Block 3: Alphonsus Crater impact, oblique geomorphology	None

Fig. 3.5 Ranger Block 3 (7, 8, and 9) TV cameras. Wide Angle cameras located in center in both banks. (Credit: NASA/JPL-Caltech)

provide radiation measurements (Geiger counter, ion chamber, dosimeter), due to perceived lack of resources (time, personnel) for required integration. As political pressure mounted, the emphasis for Ranger became preparation for Apollo landings and not science. Blocks 4 and 5 were to have far more redundancy, extensive overhaul in the identification and replacement of faulty components, a larger stage for midcourse correction, as well as a larger payload with more emphasis on planetary science, but both were cancelled, as budgets were cut and the Lunar Orbiter started to provide imaging data, as described below. The successful Rangers confirmed the presence of a modestly debris covered lunar regolith that was compacted enough to support landed hardware. Rangers didn't actually measure bearing strength: that data was provided by Lunar Surveyor. In close-up images the surface was relatively flat with 'dimple' craters.

Rangers provided the basis for the Mariner missions, deep spacecraft buses providing support for mission operations, and imaging systems in space. Throughout its lifetime, the Ranger Program faced development risks due to new environmental demands (deep space vacuum, weightlessness, radiation and thermal extremes) or changing hardware constraints (e.g., switching from Atlas Vega to the larger Air Force Atlas Agena launch vehicles) as well as shifting and complex management jurisdiction issues aggravated by 'management by committee'. The result was persistent turmoil, shifts in engineering and science priorities, great changes in design of major components and mission concept and in personnel. Weight and volume constraints became more stringent, and thus payload capability and redundancy were reduced, but then returned to previous less stringent levels when it was too late in the schedule to return to the original more robust design. Meanwhile, major conflicts arose within the science community, in regard to priorities for 'sky' as opposed to 'surface' science, and within the aerospace community, in regard to civilian versus military control and management style (loose academic versus tight industrial). These conditions resulted in a pattern of major setbacks and redesigns, as well as increased schedules and costs. Close-up imaging of the surface became the primary and then the only 'science' focus and set the stage for the major role and importance of imaging on future missions. Despite failures affecting the first six Ranger probes, after persistent efforts, and two reorganizations of management, three Rangers ultimately achieved mission goals. Most importantly, the stage was set for planetary science, especially in regard to the essential need for samples for long study. Ranger data provided the basis for 3D models of regolith and ongoing steady state processes of regolith development. From an engineering standpoint, the Ranger Program provided the basis for deep spacecraft tracking and communication, 3-axis control, autonomy, directional control, course corrections and landing maneuvers, as well as steerable antenna, and digital signal processing.

In subsequent programs, NASA used its own launch vehicles to avoid some of Ranger's problems. JPL, which had managed the Ranger program while simultaneously working on the very successful Surveyor program, had relatively little involvement with the Moon for decades after these two programs were completed.

3.3 The Lunar Orbiter Program

In contrast, Lunar Orbiter [5, 8–11] (Figs. 3.6, 3.7, 3.8, and 3.9), originally considered lower priority (to the Apollo program) than the Ranger and Surveyor programs, became the poster child for a 'successful project'. Lunar

Fig. 3.6 Top: Lunar Orbiter Spacecraft. Bottom: Lunar Orbiter Camera System [10]. (Credits: NASA/NSSDC)

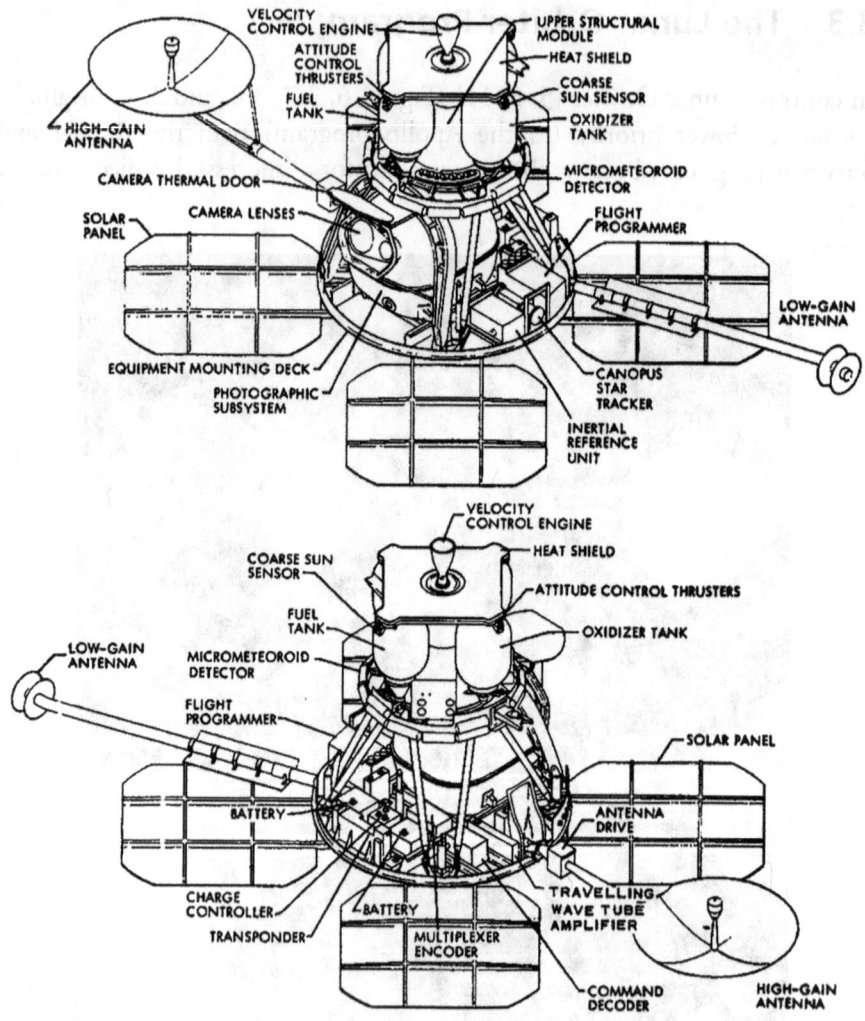

Fig. 3.7 Labelled Schematic of Lunar Orbiter Spacecraft in two orientations. (Credit: NASA)

Orbiter was a second-generation spacecraft. NASA/Langley took on the project management of Orbiter after Ranger and Surveyor, both JPL projects, were already underway. In fact, the Lunar Orbiter, originally conceived as a spinner, grew out of the JPL concept for a Lunar Surveyor orbital component. The two JPL projects were using up the available resources there, as JPL engineers struggled with and ultimately overcame technological challenges that would lay the foundation for its success in future deep space missions. JPL

Fig. 3.8 Top: Iconic oblique view of Crater Copernicus by Lunar Orbiter 2. Bottom: Lunar Reconnaissance Orbiter Map View of Copernicus for Comparison. (Credit: NASA)

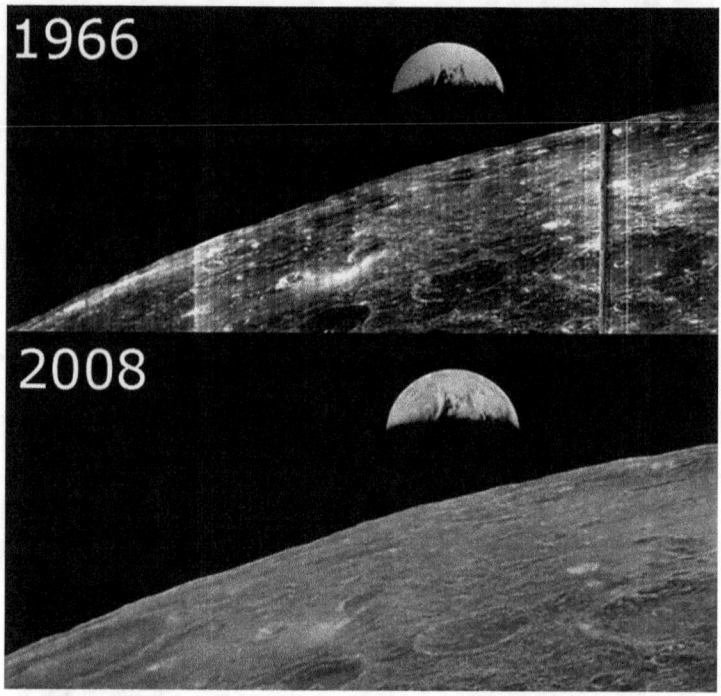

Fig. 3.9 Iconic first view of Earth from the Moon by Lunar Orbiter 1, August 23, 1966. Top is original image. Bottom is processed by Lunar Orbiter Imager Recovery Project (LOIRP) [12]. (Credit: NASA)

was still responsible for the trajectory design, navigation and tracking, and operation. Boeing was given the responsibility for attitude control systems including the inertial reference unit, sun sensor, canopus sensor, and reaction control systems (e.g., thrusters) crucial for orientation of the photographic system. The photosystem was thermally and structurally isolated from the rest of the spacecraft in a 'bathtub' configuration to mitigate the impact of extreme thermal cycling in lunar orbit. Problems with the electrical system when the spacecraft was not in solar power mode were eliminated by changing the orbit. Redesign of fuel and oxidizer tanks (with heavier gauge material) was required to eliminate initial issues with rupturing, and redesign of the attitude control system was required to reduce mass.

Major challenges were the short time frame and limited budget available, resulting in concurrent development and testing. Many of the 'lessons learned' from Ranger could be applied to Orbiter, including far better definition and control of project activities and costs from the beginning. In fact, Orbiter was the first NASA project to include cost, delivery schedule, and technical performance incentives as an explicit part of the contract agreement from the start. A new approach was used for project management: a major role for industry in making components for minimal resources with cost incentive contracts, as well as concurrent development and testing. Thus, Lunar Orbiter managed to stay within the cost and time cap with relatively small penalties. The basis for the NASA mission life cycle model was created in the Lunar Orbiter program, including Preliminary Design Review (PDR) to discover design issues, followed by Critical Design Review (CDR) to confirm issues were resolved, followed by building and formal NASA acceptance reviews with tests in three stages: qualification, endurance limits, and subsystem integration.

Orbiter imaging was designed to meet Apollo landing site selection requirements of 0.5 m spatial resolution for equatorial potential landing sites from a 40 km altitude orbit (25 m resolution over larger areas). Low inclination orbits were utilized to minimize orbital instabilities. Ultimately, existing Boeing 3-axis stabilized spacecraft and a scaled-down version of the Kodak-produced Air Force photosystem provided the basic hardware for Lunar Orbiter, though modified methods for image capture, remote production, and transmission had to be invented. The low speed film used was less susceptible to radiation damage.

The fascinating specific methodology was (quote Bruce Byers [8]):

…Film from a supply reel passed through a focal plane optical imaging system, and controlled exposures were made. Once past the shutter, the film underwent a semi-dry chemical developing process and then entered a storage chamber. From here it could be extracted upon command from the ground for scanning by a flying-spot scanner and then passed on to a take-up reel.

The line-scanning device consisted of a cathode-ray tube with a rotating anode having a high-intensity spot of light. The scanner optics of the moving lens system reduced by 22 times this point of light, focused it on the film transparencies and scanned them. A photomultiplier then converted the light passing from the scanner through the film into an electrical signal whose strength would vary with the density of the emulsion layer of the film. This signal would then be transmitted to a receiving station on Earth and reconstructed. The Eastman Kodak Company would upgrade the system for the demands of the Boeing orbiter and its mission.

A significant part of the improvement in the system was the introduction of the Kodak Bimat process, which eliminated the necessity to use "wet" chemicals on the film. Instead, a film-like processing material was briefly laminated to the exposed film to develop and fix the negative image and, if the need existed, to produce a positive image. In the case of the Boeing orbiter this 3.8 second step was not used, and only negatives were made. Once the film had been developed and fixed, the Bimat material separated from the film and wound onto a storage spool.

Although all other telemetry data were encoded and sent on carrier subfrequency, the photographic film was exposed, and the photo data not encoded but transmitted as frequency modulated analog signals to be received, captured on photographic recording equipment that was used to generate high resolution film 'originals' back on Earth. The originals of Lunar Orbiter data here on Earth are in the form of photographic film. The first automated process was developed to deal with flight telemetry: data were recorded on magnetic tape allowing higher data processing speed, with systematic procedures which included tape backups, tape changing, validation of processed data, and storage.

The future landing site selection committee set Lunar Orbiter trajectory and operational requirements. The last two orbiters flown had polar orbits for overview of major terrains and features (mare, highland, craters)) for most of the lunar surface, whereas the first three orbiters had equatorial orbits. Contiguous high resolution photo coverage of multiple potential landing site areas screened a variety of terrains on the basis of local relief and identified 'smooth ground' for landing Surveyor as well as Apollo. The committee had to contend with Lunar Orbiter photo readout limitations including (1) 192 pictures total for mission, (2) no picture taking during readout and processing, (3) readout in 1 (lowest resolution), 4, 8, and 16 (highest resolution) overlapping frames at a time, and (4) one such readout, or pass, allowed per orbit.

Geologist Hal Masursky of the USGS provided the methodology for structural and stratigraphic mapping already developed for Earth studies. This approach was further augmented by inclusion of recent information on lunar crater densities, surface roughness, and albedo from Ranger pictures.

Further Apollo requirements for radiation flux measurements, high energy solar events, and meteorite flux over a 2-week period, as well as the lunar gravitation field and figure resulted in the addition of several instruments, including radiation and micrometeoroid monitors, and the use of navigation and tracking instruments for gravity field and figure measurements in Block one orbiters. The discovery of mascons (areas of high mass concentration) from these measurements led to the understanding that more tracking data was needed before lunar landings. Block two orbiters, which never flew, were to carry further scientific instruments, including gamma-ray, X-ray, and visible/near IR spectrometers, radiometers, bistatic radar, magnetometers, solar plasma (electrons) and ion spectrometers on 1-year missions. The orbital trajectories of Block 2 were eventually utilized by the Apollo service modules, which carried various combinations of these and other instruments as part of the Scientific Instruments Module (SIM) bay.

All five orbiters collected data (Table 3.3), with minor problems during the missions. Lunar Orbiters provided heritage for all future orbiters. The next block was to carry more comprehensive science payload. Ironically, that block of lunar orbiters was cancelled due to work force shortages.

Orbiter 1: Though the spacecraft experienced manageable problems with the thermal protection system and battery degradation, both fixed on subsequent missions, the photosystem performed 'with the precision of a swiss watch' and the film moving system 'operated like a thrashing machine' [8]. Two photosequences, a sixteen-frame high-resolution mode and a four frame medium-resolution mode, were made of the Mare Smythii region. A problem with synchronization of the V/H sensor and lens (solved in later missions) resulted in smearing of the high resolution pictures, and film imperfections resulted in some density variations, but the first pictures of terminators, the Earth from the Moon, the farside, and the first oblique photographs were provided.

Orbiter 2: This orbiter focused on determining the 'smoothness' of potential landing sites for Apollo landers, the remaining Rangers, and Surveyors, with focus on dealing with the apparently conflicting data from Ranger 7 and Lunar Orbiter 1 as to how rough the ubiquitous crater rays were. Iconic views of Copernicus Basin, Marius Hills, and the Reiner Gamma magnetic swirl anomaly were provided. Reseau marks were added to photos to deal with nonlinearity. The film advance mechanism failed near the end of the mission.

Table 3.3 Lunar orbiter missions

#	Taking data	Goal	Issues
1	8/18-29/66	Apollo Landing Site Survey	Manageable thermal and power system issues, redesigned for future missions
2	11/18-25/66	Apollo Landing Site Survey	Film advance failed near mission end
3	2/15-23/67	Apollo Landing Site Survey	Film advance problematic
4	5/11-26/67	Mapping Nearside, polar orbit	Film issue from solar flare
5	8/6-18/67	Mapping Farside, hi-res survey, stereo coverage, first test DSN, polar orbit	None

Orbiter 3: In a somewhat higher inclination orbit, this orbiter was launched while Orbiter 2 was still in orbit, confirming advanced DSN tracking capabilities required for Apollo. Apollo and Surveyor potential landing sites were confirmed. The film advance motor was problematic.

Orbiter 4: The first polar orbiter demonstrated the flight programmer would work on the farside. Solar Flare events and a high radiation dose from the Van Allen radiation belts caused a series of problems, including partial film fogging, temperature control issues, and a noisy encoder. Coverage was provided of 99% of the nearside with ten times better resolution than available from the Earth, and 80% of the farside. The orbit was raised to gather more gravity data.

Orbiter 5: The second polar orbiter provided the iconic picture of the full Earth, and a realization of the potential for remote Earth viewing. Particularly relevant were the higher resolution and greater coverage stereo photos, including near-vertical convergent telephoto stereo for Apollo landing sites, a broad survey of previously unphotographed areas, coverage for additional surveyor landing sites, and closer looks at Apollo landing sites near the terminator at low elevation to determine relief. More gravity, micrometeoroid and radiation data were also provided. The first test of the manned space flight network was successfully performed.

Of specific interest to those who would select and analyze Apollo landing sites were the determination of (quote Bruce Byers [8]):

> the freshness of the features in the site. Earlier Orbiter missions have shown emphatically that most lunar terrain has a subdued appearance at all Orbiter scales so that little new is learned from high resolution photography. Fresh young craters (mostly light) and fresh young rock units (mostly dark) that are not yet much modified by repeated cratering and wasting potentially reveal the most about rock type and origin, both in photographs and when sampled on the ground. Old terrains show effects of the processes that waste lunar slopes and though these are of interest, they seem to be sufficiently sampled in high resolution photography by earlier Orbiter missions, except for very high and steep slopes. A few high and steep slopes and other non-fresh targets have been selected for the purpose of rounding out terrain sampling.

Orbiter accomplishments were somewhat eclipsed by the Apollo missions, and program influences were somewhat minimized by the simultaneous planning for the Apollo program before the survey of the lunar surface was completed. No orbiter missions followed 5, due to manpower shortages, missions which were to have provided higher resolution coverage overall. However, the Orbiter series provided the basis for (1) selecting landing sites and providing landing ellipses (avoiding craters and younger terrain), (2) determining primary and secondary micrometeorite impact was no particular threat, and (3) designing radiation shielding for average short-term exposure. The Orbiter program provided the heritage for future orbiters (e.g., Viking), laid the foundation for cooperation between the manned and unmanned program components and provided crucial experience in mission operations for Apollo orbiters.

3.4 The Lunar Surveyor Program

The Surveyor lander series [5, 11, 13–15] (Figs. 3.10, 3.11, 3.12, 3.13, 3.14, and 3.15) were the first American soft-landing probes on the lunar surface. The Surveyors played an important role in the success of the Apollo project, confirming that the regolith would support a lander. One well known National Academy scientist (Tommy Gold) hypothesized that the regolith would act as quicksand ('Gold' Dust) and a lander would sink and be swallowed up. That was one reason that large footpads were used. In actuality, depth of footpad penetration was observed to range from 3 to 12 to centimeters. The Surveyor engines were cut at 3.5 m above the surface for a 'free fall' landing. The 1000 kg landers were 3.3 m high by 4.5 m in diameter, and their aluminum tubing tripod structures provided excellent mounting structures for payload components. The landing gear consisted of, in order, crushable footpads, mounted on three legs with shock absorbers and upwardly rotatable along hinged axes, as well as three cylindrical crushable blocks mounted under the frame near the leg attachments. After landing legs reextended to their original length. Surveyor and Orbiter were originally conceived as one program, analogous to what became the Viking mission to Mars. In any case, Surveyor became the 'follow on' soft lander to the Ranger hard lander. JPL could take direct advantage of developments that had been necessary for the hard lander. The biggest challenges were developing the technologies needed for closed-loop, radar-controlled landings and the world's first hydrogen-oxygen rocket stage, the Centaur. Also, Surveyor would require the integration of the most complex payload to date. Despite the difficulties of these tasks, Surveyor was largely successful: all but two of the seven surveyors landed and sent back data from progressively more complex payloads. The Surveyor project provided in situ 'ground truth' for potential Apollo landing sites, visiting the most likely equatorial maria sites as well as the prominent, young crater Tycho on the last mission. Surveyor landers sent back thousands of images (via vidicon TV camera with photometric targets), as well as other science data (Table 3.4),

Fig. 3.10 Panoramic mosaics of Survey 6 (top) and Survey 7 (bottom) landing sites. (Credit: NASA)

including measurements of soil chemistry (via alpha-scattering spectrometer) on all but the first two missions and measurements of mechanical properties (via gauges on a highly motorized scoop for sample collection) on all but two of the missions. The last three Surveyors used footpad magnets to determine magnetic susceptibility of lunar regolith. Ultimately, the Surveyor program returned over 88,000 high resolution photographs in situ on the lunar surface. These measurements were essential for designing and planning astronaut surface activities. They Surveyor Program provided heritage for future landers.

All but one of the Surveyors landed in mare equatorial areas, which were considered the most likely Apollo landing sites. The final surveyor landed close to the fresh, young crater Tycho, in order to sample lunar highlands, the most prominent terrain on the moon. (Table 3.4). Selenographic coordinates were determined by matching features observed surrounding the Surveyor lander with corresponding features on lunar orbiter photographs. The later Surveyors survived lunar nights, taking data during more than one period of

Fig. 3.11 Surveyor 6 Footpad 2 (NASA) showing rocky portion of surrounding regolith. (Credit: NASA)

illumination. All Surveyor landing sites were selected because they were being considered for Apollo landing sites.

Surveyor I. Most importantly, established that we could land safely on the Moon. It landed within 15 km of the target, on the flat portion of a 100-km crater (Flamsteed) inside of a small (200-m) crater in southwestern Oceanus Procellarum in the western hemisphere near the equator.

Surveyor II failed to complete a midcourse correction successfully when one of the thrusters failed to ignite and the spacecraft tumbled due to unbalanced thrust. Instead of landing at Sinus Medii as planned, it crashed near Copernicus.

Surveyor III, despite failure of engines to shut down as designed, landed successfully in a 200-m crater southeastern Oceanus Procellarum that

Fig. 3.12 Surveyor 1 shadow. (Credit: NASA)

eventually became the Apollo 12 landing site. It had far more instrumentation than Surveyor I, including an active sampling system capable of providing information on the mechanical and structural properties of lunar regolith.

Surveyor IV had the same payload as Surveyor III and may have landed, but radio contact was lost 2.5 minutes before the planned touchdown at Sinus Medii, and was never reestablished.

Surveyor V landed on a steep inner wall slope of a 10-meter scale crater (part of a crater chain) in Mare Tranquillitatis. An alpha-backscatter instrument was added, replacing the surface sampler, to take compositional measurements. This Surveyor survived lunar night and was able to resume operation when turned back on during the day.

Surveyor VI, with a payload similar to Surveyor V, performed better than any of the previous landers. It landed in a flat heavily cratered area 200 meters west of a 30-m mare ridge in Sinus Medii in the center of the nearside, and purposely performed a 2.5 meter 'hop' maneuver from the original landing

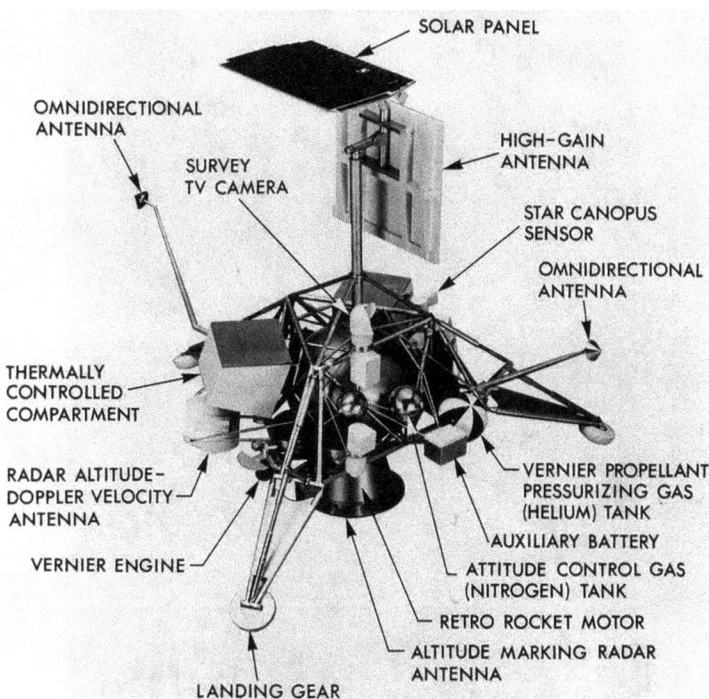

Fig. 3.13 Surveyor Lander Schematic. (Credit: NASA)

site to provide a view of regolith disturbed by landing thrusters and better understanding of soil mechanics.

Surveyor VII, carrying the most extensive payload, landed two miles from the target on a rocky ejecta blanket near the north rim of the fresh, young crater Tycho, terrain considered the most hazardous of any selected landing site. It operated over two lunar days. The surface sampler was used to place the alpha-scatter experiment, which failed to fully deploy, on the surface, subsequently moving it two more times. The lander also performed star surveys and took pictures of the Earth and the solar corona.

More extensive science investigation was one of goals of the Surveyor program, thus the Surveyor lander had more ambitious scientific payloads than other pre-Apollo programs. The onboard sensors generated more wide-ranging data from several instruments and multi-functional onboard sensors representing a broad cross-section of disciplines and multi-disciplinary investigations. Thus, the Surveyor Program created the basis for the approach for analysis, publication, and sharing of science data for future NASA missions. In addition, the Surveyor program laid the groundwork for a most important

Fig. 3.14 Surveyor 3 picture taken by Apollo 12 astronaut (top) and with Apollo 12 astronaut nearby and Landing Module in background. (Credit: NASA)

capability for scientific analysis as well surface exploration in general. A capability that was next utilized in the ALSEP program: surviving lunar night. This capability is challenging even to this day, five decades later!

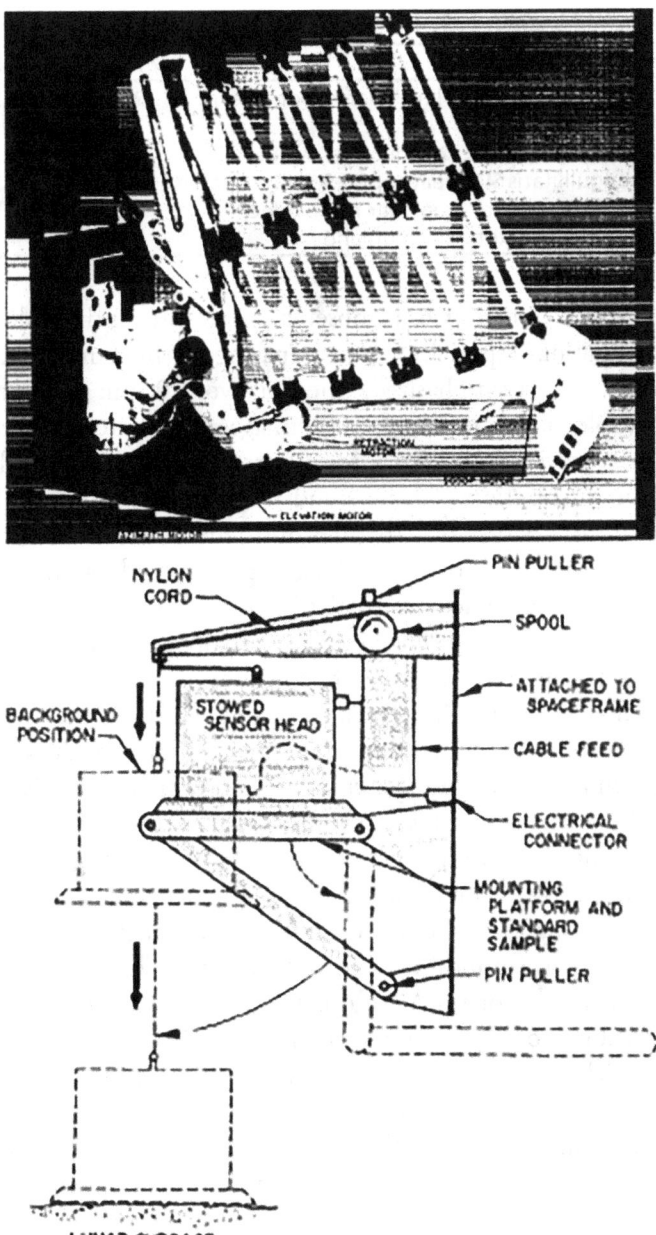

Fig. 3.15 Surveyor Scooping arm (top) and Alpha backscatter instrument (bottom) deployment mechanisms. (Credit: NASA)

Table 3.4 Successful lunar surveyor landers

#	Location and description	Lat	Long
1	Oceanus Procellarum SW, level mare floor level of Flamsteed Ring	2.45 S	43.2 W
3	Oceanus Procellarum SE, 200 m crater wall	2.97 S	23.3 W
5	Mare Tranquillitatis SW, near top of 9 × 12 m crater	1.50 N	23.2 E
6	Sinus Medii, level mare area near mare ridge	0.53 N	1.40 W
7	Tycho N ejecta blanket, hilly highland	40.86 S	11.47 W

The importance of taking visual observations in situ, at the scale of human or rover exploration, especially when preparing for humans to walk on the Moon, cannot be overemphasized. The Surveyor 70 mm television camera fulfilled that role, obtaining over 86,000 images with up to 1 mm resolution. These images provided the basis for understanding the lunar landscape, including the distribution of craters and rocks in terms of size, shape, and frequency, and the nature of structures and relief on tens of cm to tens of meter scales, information critical for designing rovers and training humans for traverses of the lunar surface. Of particular note were the many small nearly rimless craters, some dimple-shaped, difficult to observe under high solar illumination angles, and irregular debris filled craters thought to be produced by secondary impacts, accounting for most of the relief, and the association of higher populations of brighter, rocky fragments with craters, and blockier fragments with larger and younger craters. The size-frequency distribution of these ubiquitous smaller craters appears to indicate a steady-state production through ongoing meteoroid bombardment. Surveyor images firmly established cratering as the major influence on terrain formation at every scale and provided supporting evidence on soil mechanical, magnetic, and structural properties, as well as composition when the cameras were used to document visually the operations or results of other sensors described below. Based on the observed ability of the fine-grained regolith material to preserve cratering imprints, particle size was estimated to be between 2 and 60 μm. Variations in albedo with depth were also observed, with brighter material, from more recent ejecta, being present on the surface and darker material several mm below on Surveyors III, V, and VI, all of which landed within or close to small, relatively recent craters.

Mapping of the landing sites from camera images required the use of several techniques. These included [13, 14]:

(a) focus ranging: a series of narrow-angle near-field images is taken at a range of focus settings at each camera elevation position at that azimuth (direction in a 360 circle around the lander). The most 'in focus' views of fea-

tures at given elevation and azimuth are then used as control points to make equal elevation/equal azimuth contour maps.
(b) stereogrammetry: a series of adjacent, largely overlapping images of the surrounding area are used to produce 3D projections of the local terrain.
(c) shadow measurement: spacecraft profile along east-west line using shadow projections of spacecraft features of known dimensions, and knowledge of solar elevation and azimuth are used to determine distance and dimensions of features of surrounding landscape along this line.
(d) correlation: Features in Lunar Orbiter features are correlated with features in surveyor field of view to make small-scale far field maps. From these correlations, topographic maps were made

Instruments that provided direct information on mechanical, structural, and magnetic properties of regolith included strain gauges mounted on landing gear, magnets mounted on the footpads, and surface samplers which acted on trenching tools on two of the landers and allowed bearing and impact tests on lunar soil. These tests indicated that the lunar regolith is compressible, at least somewhat cohesive, and has static bearing increasing with depth. The adhesion of soil to spacecraft surfaces was observed, as was evidence of 'sandblasting' of spacecraft surface by particles ejected during blasts of the propulsion system. However, the truly 'abrasive' quality of lunar regolith was fully evident during the Apollo landings when 'brushing' was used to remove dust and abraded the pressure suits. Consistently, the high polarization of light scattered from rock fragments observed on Surveyor 7 suggests glassy surfaces. As described above, the operation of the propulsion and attitude control system after landing to lander 'hopping' also provided useful data on the nature of fine particles or dust in the regolith, in particular establishing how landers would disturb surrounding surface regolith. In vertical firing tests, erosion of a layer of up to 1 cm of soil extended to 2 meters, and fragments of up to 4 cm were displaced up to 20 cm. In addition, gases diffused into the soil caused a 20 cm crater about 1 cm deep to form. The horizontal component of ejection direction depended on the direction of engine firing, with offsets preferentially in the direction of firing. Variations in magnetic susceptibility in lunar soils could imply variation in native magnetite content (ferromagnetic material). Such material could also by introduced by frequent meteorite bombardment of the Moon. A magnet was mounted on one or more lander footpads (and in some cases on the surface sampler) to determine the magnetic susceptibility, and by implication ferromagnetic content, of lunar soils, supplementing the alpha-backscatter experiment. The amount of attracted lunar dust was used to estimate the amount of ferromagnetic material present in lunar soil and infer overall composition. A small amount of magnetic material was found at all sites, but no significant differences were found between sites, implying that the ferromagnetic material is of meteoritic origin.

The alpha-backscatter experiment provided the first chemical analyses of lunar surface material, estimating elemental abundances for several samples in the later Surveyor missions. The Alpha Backscatter instrument was deployed as shown in (Fig. 3.15). Measurements were taken of an onboard standard sample to provide calibration for in situ surface samples. Data accumulation times varied from 7.5 to 72 h. Surveyors V analyzed two lunar samples, Surveyor VI 1, and Surveyor VII 4. Lack of indications of the presence of either C or N minimized the possibility of contamination of the samples with fuel from the lander system. In addition, measurements appear to have been made on undisturbed soil sites. Measurements for all sites indicate, in order of abundances major elements Oxygen, Silicon, and Aluminum, just as on Earth. The mare sites appear to be analogous to terrestrial basalts. The highland site has half as much of the heavier (iron-group) elements. These measurements ruled out a meteoritic origin for the Moon, clearly demonstrating that the Moon is NOT primordial material, a major hypothesis at that time. The atomic fraction of oxygen (>0.5) indicates that all metals are present as oxides, as in terrestrial silicates, and spectra indicate the samples consist mostly of a mixture of feldspar and pyroxene, consistent with terrestrial igneous rocks.

Surprisingly, considering how challenging the lunar thermal environment is, Surveyor had no instruments specifically designated to measure thermal conditions. Temperature estimates were made using a Lambertian model for time of day versus temperature based on nearside lunar isothermal contours derived during totality of a 1964 solar eclipse. However, the lunar surface is not Lambertian (brightness temperature is not constant for different angles of observation), so corrections were based on observations done by [16] at a 90° angle, as well as [17] (at 30° and 60° angles). Thermal measurements were made in April of 1967 while an eclipse occurred during the Surveyor III landing. During the missions, several temperature sensors on external components of the landers (solar panels, planar array, electronics) provided a basis for correcting for local conditions and understanding the cyclical extremes in the local thermal environment of the lunar surface. Corrections in surface temperatures for local conditions were found to be most correlated with sun elevation angle relative to local lunar surface slope, which varied from lander to lander as spacecraft orientation and surface tilt varied. Consistent variations in temperatures predicted based on eclipse and in situ observations were thought to be due to variations in thermal conductivity and density with soil depth [18].

3.5 Pre-landing Crewed Apollo Missions

Pre-landing Crewed Apollo Missions performed a crucial role. They demonstrated hardware and crew capabilities required for the Apollo landed missions, including performing critical burns, achieving lunar orbit, performing separation, rendezvous, docking, as well as simultaneous and separate operation of the three connected and deployable components: lunar landing module, service module, and Earth landing capsule. They also demonstrated the Apollo life support system in space and on the Moon, and confirmed communication and tracking capability of the Apollo spacecraft. In this way, the pre-landing Apollo missions themselves established a number of 'firsts' [19, 20].

Apollo crewed missions began with Apollo 7. Apollo 2 and 3, planned uncrewed ground tests, were retroactively renamed to AS-202 and AS-203 after the fire in the Apollo 1 command module during a launch rehearsal test. Apollo 4 through 6 were flown as uncrewed tests, 4 and 6 as uncrewed tests for the Saturn V, and 5 as an uncrewed test for the Lunar Module.

From 1968 to 1969, prior to lunar landing, Apollo crewed missions demonstrated crucial capabilities including

- operation of the command module in Earth orbit (Apollo 7)
- operation of the command module to, from and around the Moon (Apollo 8)
- operation of the full Apollo configuration, both command and lunar module, in Earth orbit (Apollo 9)
- all operations for lunar landing at the moon except actually landing (Apollo 10).

Apollo 7 [21] (Figs. 3.16, 3.17, and 3.18) was the first crewed flight in the Apollo program with the principal goals of demonstrating Apollo rendezvous capability (CSM with spent S-IVB), flight crew performance on the command and service module and ground crew performance in operations support facilities during a mission, and the first live television broadcast from space. Minor issues were lack of full deployment of the lunar module adaptor panels, a problem solved by the use of explosive jettisoning in later flights; inadequately cured window sealant resulting in minor window fogging (resolved by Apollo 9); as well as underperformance of battery chargers, and occasional overheating of one of the three fuel cells (both power system problems solved for subsequent missions).

Fig. 3.16 Top: Apollo 7 Crew in first live Television broadcast from space. Bottom: Mission Control Houston during Apollo 10 mission. (Credits: NASA)

Apollo 8 [22] (Figs. 3.19 and 3.20) was the first human mission to the Moon with the principal and challenging goal of demonstrating the capability for getting the astronauts to and from the Moon and a 'poster child' for the technological success and human ingenuity of the Apollo program. Its technical achievements were successful completion of translunar injection and mid-course correction maneuvers, ground and crew coordination of communication and navigation activities, and demonstration of thermal control and life

3 The Context: Robotic Precursors Ranger, Surveyor, Orbiter

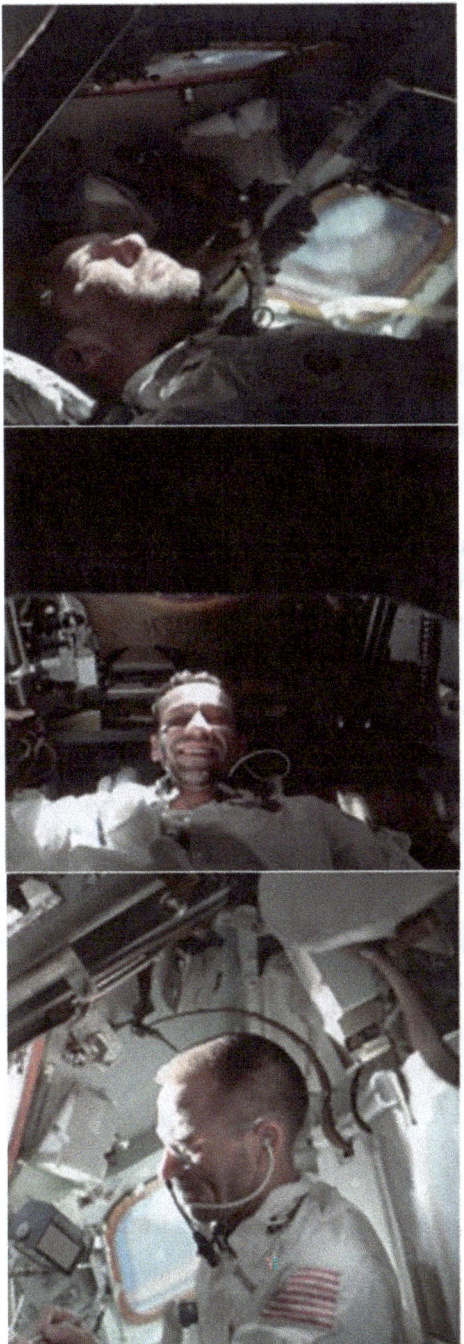

Fig. 3.17 From Top to Bottom Apollo 10 astronauts Schirra, Eisele, and Cunningham in orbit. (Credit: NASA)

Fig. 3.18 From Top: Schirra, Eisele, Cunningham after Apollo 7 splashdown on deck of Essex. Bottom: Kranz, Lunney, and Griffin celebrating in mission control after Apollo 7 splashdown. (Credits: NASA)

Fig. 3.19 From Top: Marilyn Lovell (center) accompanied by Alan Shepard (left) in Mission Control Visitors' Gallery. Bottom: Apollo 8 capsule photographed at 40,000 feet during reentry by support crew. (Credits: NASA)

Fig. 3.20 Iconic Earthrise over Moon taken by Bill Anders during lunar obit of Apollo 8. (Credits: NASA)

support systems in the command and service module in the deep space and lunar environments. Only one out of the three originally planned midcourse corrections was required. The return to Earth, via a double skip maneuver, was flawless and exactly on target. Voice transmission, which included a timely telecast with readings from Genesis on Christmas Eve, were of excellent quality, the best ever achieved from space.

Apollo 9 [23] (Figs. 3.21 and 3.22) was the first flight demonstration, including completely independent operation, of the crewed Lunar Module. Tests included complete checkout of all the systems of the Command, Service, and Lunar Modules separately, and while all were linked or in all anticipated docking configurations. Successful demonstration of docking and undocking of the Command and Lunar Modules under simulated Apollo landing mission conditions was the top priority. The Mission also demonstrated the use of the astronaut Portable Life Support System (PLSS Backpack) for the first time in deep space. "Apollo 9's success ensured that the next Apollo mission would go to the Moon [24].

Fig. 3.21 Top: Astronaut Schweikart on Lunar Module Spider's front porch evaluating PLSS backpack during Apollo 9 Earth orbit. Bottom: Astronaut Scott supporting the undocking of Spider from Service Module Gumdrop during Apollo 9 Earth orbit. (Credits: NASA)

Fig. 3.22 Apollo 9 Lunar Module flown by astronaut Schweickart in Earth Orbit. (Credits: NASA)

Apollo 10 [25] (Figs. 3.23 and 3.24) was the first rehearsal of all of the modules in the lunar environment, including all operations except the final touchdown and take-off from the lunar surface. Of particular importance were the data gathered during the rehearsal of the landing on the lunar gravitational effects, Earth-based tracking techniques for the LM, and checking out of LM programmed trajectories, use of radar, and flight control systems. The first color pictures of the Moon were transmitted to the Earth. After the landing simulation, which resulted in a minor problem involving gyration of the ascent stage due to an error in the flight plan (corrected in subsequent missions), the crew performed landmark photography and tracking.

The precursor programs and missions in this chapter, including Ranger, Lunar Orbiter, Lunar Surveyor, and the pre-landing Apollo missions, were essentially early components of the Apollo program. The nature and timing of each was systematic and they were designed to prove operational capabilities needed and to establish the nature of the lunar environment and demonstrate

3 The Context: Robotic Precursors Ranger, Surveyor, Orbiter   95

Fig. 3.23 Top: Apollo 10 on crawler between VAB and launchpad. Bottom: Apollo 10 Lunar Module Snoopy being tested in Lunar Orbit by Stafford and Cernan. (Credits: NASA)

approaches to mitigating risks (thermal and radiation extremes, rocky and rugged cratered terrain with ubiquitous fine ejecta of initially questioned ability to support humans or hardware. Earlier missions, especially Lunar Orbiter, were also essential for finding and characterizing potential landing sites. Without these earlier missions, the Apollo landings themselves would not have been possible.

Fig. 3.24 Support Crew inspects egress slide wire gondola before Apollo 10 mission. (Credit: NASA)

References

1. NASA/JSC, Apollo Program Summary Report: Synopsis of the Apollo Program and Technology for Lunar Exploration (NASA-TM-X-68725, 1975. https://ntrs.nasa.gov/api/citations/19750013242/downloads/19750013242.pdf
2. Newell, Homer, Beyond the Atmosphere: Early Years in Space Science (NASA SP-4211), GPO, 1980.
3. Bilstein, Roger, Orders of Magnitude: A History of NACA and NASA, 1915-1990 (NASA SP-4406,1989), NASA, 132 p., 2011.
4. Portree, David, NASA's Origins and the Dawn of the Space Age (Monographs in Aerospace History #10, 1998), Military Bookshop, 134 p., 2011
5. Dick, Steve, Stephen Garber, Jane Odom, Research in NASA History, CreateSpace Independent Publishing Platform, 86 p., 2009.

6. Hall, Cargill, Lunar Impact: A History of Project Ranger (NASA SP-4210), NASA, 564 p., 2011.
7. NSSDC, Ranger Missions, Experiments, and Data, https://nssdc.gsfc.nasa.gov/planetary/lunar/ranger.html. last visited 10/01/24.
8. Byers, Bruce, Destination Moon: A History of the Lunar Orbiter Program (NASA TM X- 3487, 1977), NASA, 220 p., 2011.
9. Bowker, David E. and Kenrick B., Lunar Orbiter Photographic Atlas of the Moon (NASA SP- 206), NASA, 41 p., 1971.
10. NSSDC, Lunar Orbiter Missions and Data, https://nssdc.gsfc.nasa.gov/planetary/lunar/lunarorb.html. last visited 10/01/24
11. Kloman, Erasmus, Unmanned Space Project Management Surveyor and Lunar Orbiter (NASA-SP-4901, 1997), Bibliogov, 54 p., 2013.
12. Lunar Orbiter Imaging Recovery Project (LOIRP). LOIRP modification at PDS Imaging Node. https://planetarydata.jpl.nasa.gov/img/data/lo/LO-1001
13. Godwin, Robert, Surveyor Lunar Exploration Program: The NASA Mission Reports, Apogee Books, 178 p., 2007.
14. NASA, Surveyor: Program Results Final Report (NASA-SP-184, 1969), 436 p., 2013.
15. NSSDC, Surveyor Missions, Experiments and Data. https://nssdc.gsfc.nasa.gov/planetary/lunar/surveyor.html. last visited 10/01/24.
16. Sinton, W.M., Thermophysical Behavior of the Lunar Surface, In Kopal, Z., Ed., Physics and Astronomy of the Moon. Academic Press, New York, 1962.
17. Shorthill, R. and J.M. Saari, Infrared observation on the eclipsed Moon, Adv Astron Astrophys, 9, 149-201, 1972.
18. Winter, D.F. and J.M. Saari. A particulate thermal model of the lunar soil. Astrophys J., 156, 1135-1151, 1969.
19. Harland, David, NASA's Moon Program: Paving the Way for Apollo 11, Springer, 472 p., 2009.
20. Lund, Tom, Early Exploration of the Moon: Ranger to Apollo, Luna to Lunniy Korabl, Springer, 410 p., 2018.
21. Apollo 7 Overview, Lunar and Planetary Institute, Resources. http://lpi.usra.edu/lunar/missions/apollo/apollo_7. visited 10/1/24.
22. Apollo 8 Overview, Lunar and Planetary Institute, Resources. http://lpi.usra.edu/lunar/missions/apollo/apollo_8. visited 10/1/24.
23. Apollo 9 Overview, Lunar and Planetary Institute, Resources. http://lpi.usra.edu/lunar/missions/apollo/apollo_9. visited 10/1/24.
24. French, F., and C. Burgess. In the Shadow of the Moon. University of Nebraska Press2010.
25. Apollo 10 Overview, Lunar and Planetary Institute, Resources. http://lpi.usra.edu/lunar/missions/apollo/apollo_10. visited 10/1/24.

4

The Preparation: Planning, Selecting, Training

4.1 Planning and Simulating

The plan for Apollo surface exploration was to use preselected landing sites and traverse routes. Selected areas and traverses had to meet safety criteria, minimizing required time, distance, and relief along the route, while allowing access to geological features of greatest interest in a landing site area. (Details of the scientific rationale for the selection of each site will be discussed later in the book.) 'Field stations' along the route were typically sites of surface exposures of underlying or crater-excavated rock that allowed sampling of its composition as well as characterization of local stratigraphy and geomorphological structure, and thus, potentially, determination of the site's origin. For the Apollo J mission training, terrestrial sites and routes were direct analogs in terms of distance covered and number of stations visited [1].

The astronauts had to rely on EVA simulation and training to operate 'where no man had gone before' (e.g., 4) and the innovative use of a 'back room' of co-trained geologists during their EVAs. Capturing the geological character of the site accurately and succinctly the first and only time, despite the fact that they had limited time (3 EVAs of up to 24 h in the field for the later J missions) was essential. The astronauts were restricted to tens of kilometers in the rover and normally tens of meters on foot. Science objectives would be addressed by gathering samples of rock and regolith at accessible outcrops and determining their relationship to the surrounding terrain as keys to understanding the underlying structure and stratigraphy. These constraints affected every aspect of the approach to geological field work as well as field hardware design.

To support lunar surface exploration simulation, aerial reconnaissance data with resolution comparable to Lunar Orbiter were obtained for selected terrestrial sites. Just as Orbiter data, supplemented by Apollo orbital camera coverage on some later missions, were used to derive geological unit maps for potential landing sites, aerial reconnaissance data of comparable resolution were used to derive geological unit maps for terrestrial astronaut training and mission simulation sites [1].

In the early stages of training, the astronauts were given photogeological unit maps derived as described above for field maps. They had difficulty in finding contacts for the units which had subtle slope, textural, or tonal changes challenging to observe in the field and not obvious on the basis of the simple lithological units sampled [2, 3]. Thus, the decision was made to supply the astronauts with photographic maps with traverses superposed and major landmarks identified. a limit was found to the amount of data that was digestible and useful in real time [2]. Activity checklists, mountable on the sleeve, were mnemonic or graphic and lengths were kept to a minimum. Current technology would allow access to an equivalent basemap on a handheld device with the ability to add selected details of terrain or compositional features as well as to increase resolution as approaching a site as needed.

4.2 Astronaut Selection

When NASA was established in 1958, no formal criteria for astronaut selection existed [4, 5]. Assuming a returning spacecraft would require an extended glide, as Shuttle eventually did, aviation experts assumed that aptitude for and experience in piloting jet and rocket high altitude aircraft, analogous to the X-15, would be critical skills (Randolph Air Force Base School of aviation medicine). The ability to cope with acceleration forces of up to 9g's, and deal with somewhat more complex instrumentation and control, seemed obvious additional criteria. That would have meant we should look for 'experienced pilots of high performance aircraft'.

School of Aviation Medicine at Randolph Air Force Base in Texas, published the results of an investigation regarding the selection and training of personnel for spaceflight [6]:

- the ability to deal with 'the implications of a seemingly complete break from the Earth and the protective societal matrix in a small, isolated, closely confined container with few companions'

- having 'manifest intense motivation for the project' while having a 'strong ability to cooperate to that point that they could place trust and confidence in associates and win the trust and confidence of those associates', and
- having 'positive interpersonal attitudes, mature character integration, and emotional stability involving an inner sense of duty, responsibility, self-control and restraint' and were 'adventurous but not foolhardy', while
- being 'their own instructors' but coachable, accepting additional training required in 'applied and theoretical mathematics, electronics, engineering, navigation, astronomy and astronavigation, spacecraft design and operation, space medicine, as well as training in simulators.

The independent Life Science Advisory Committee, established by NASA recommended astronauts be chosen from more 'dangerous' professions [5].

President Eisenhower had his own set of criteria [4]:

- Have a degree or the equivalent in physical science or engineering
- Be a graduate of a military test pilot school
- Have at least 1500 h flying time including a substantial amount in high performance jets
- Be younger than 40
- Be no taller than 5'11"
- Be in superb physical condition
- Possess psychological attributes specified by the Life Sciences Committee

NASA selected astronauts from 1959 to 1969 (Table 4.1) to support the human spaceflight Projects Mercury (crew of 1), Gemini (crew of 2), and Apollo (crew of 3). Each required a full backup crew complement. Pre-Apollo programs were designed to and did achieve goals essential for the success of the Apollo program. The primary goal of Project Mercury (1958–1963, 6 missions) was to send a human to and safely return a human from Earth orbit, which required creating a viable life support system as well as the capability for communication and tracking of a spacecraft in Earth orbit. The Project Gemini (1961–1966, 10 missions) goal was to develop technologies and techniques that would support Apollo operations, including longer duration flight, spacecraft rendezvous and docking, and extra-vehicular activity.

Out of almost 500 initial applicants for the first astronaut candidates, 32 were selected, after the initial review and in-depth interview process, for the strenuous physical and psychological screening process. Michael Collins [4] described this process as follows: Inconvenience is piled on top of uncertainty on top of indignity, as you are poked, prodded, pummeled, and pierced. No

Table 4.1 Astronauts selected 1959–1969

Astronaut	Org	Program
Lt. Scott Carpenter	Navy	Mercury, Gemini
Cpt. Leroy Cooper Jr	Air Force	Mercury
Lt. Col. John Glenn, Jr	Marines	Mercury
Cpt. Gus Grissom	Air Force	Mercury, Gemini, Apollo
Lt. Com. Walter Schirra	Navy	Mercury, Gemini
Lt. Com. Alan Shepard	Navy	Mercury, Apollo
Cpt. Deke Slayton	Air Force	Mercury
Lt. Col. Ed White	Air Force	Gemini, Apollo
Cpt. John Young	Navy	Gemini, Apollo
Lt. Gen. Thomas Stafford	Air Force	Gemini, Apollo
Cpt. Jim Lovell	Navy	Gemini, Apollo
Cpt. Peter Conrad	Navy	Gemini, Apollo
Col. Frank Borman	Air Force	Gemini, Apollo
Cpt. Gene Cernan	Navy	Gemini, Apollo
Col. David Scott	Air Force	Gemini, Apollo
Maj. Gen. Michael Collins	Air Force	Gemini, Apollo
Brig. Gen. James McDivitt	Air Force	Gemini, Apollo
Cpt. Richard Gordon	Navy	Gemini, Apollo
Col. Buzz Aldrin	Air Force	Gemini, Apollo
Lt. Neil Armstrong	Navy	Gemini, Apollo
Roger Chaffee	Navy	Apollo
Charles Duke	Navy	Apollo
Harrison Schmitt	Civilian	Apollo
Fred Haise	Air Force	Apollo
Alan Bean	Navy	Apollo
Bill Anders	Air Force	Apollo
Jack Swigert	Air Force	Apollo
Edgar Mitchell	Navy	Apollo
Ken Mattingly	Navy	Apollo
Alfred Worden	Air Force	Apollo
James Irwin	Air Force	Apollo
Ronald Evans	Navy	Apollo
Rusty Schweickart	Air Force	Apollo
Walter Cunningham	Marines	Apollo
Donn Eisele	Air Force	Apollo
Sturart Roosa	Air Force	Apollo

orifice is inviolate, no privacy respected …." In addition to a range of extreme physical stress tests, candidates were locked in a dark room for an extended period of time. Those who passed these tests were evaluated for their degree of interest in the program, resourcefulness, and creative 'survivor' instincts. The 'Mercury Seven' were selected from this group (Carpenter, Cooper, Glenn, Grissom, Schirra, Shepard, Slayton) in 1959 (Fig. 4.1). Two things are worth noting about astronaut selection in this time frame. Before the final selection, some candidates embraced the attitude of iconic test pilot's test pilot Chuck

Fig. 4.1 Project Mercury astronauts. top: Cooper, Carpenter, Glenn, Shepherd, Grissom, Schirra, Slayton in training; bottom: Grissom, Shepard, Carpenter, Schirra, Slayton, Glenn, Cooper with rocket. (Courtesy of NASA)

Yeager, and withdrew because they didn't consider this program significant enough to contribute to their careers.

In the early sixties, during the time of the first astronaut screenings, despite the requirement for graduation from a military test pilot school, which excluded women, who were not allowed in combat, 13 women, test pilots, many of whom had flown as Women Airforce Service Pilots (WASPS) during World War II, underwent astronaut training sponsored not be NASA but by a NASA physician using the tests and equipment he developed for NASA [7, 8] (Fig. 4.2). These women were known as the Mercury 13. He assumed that once stations were established in space, women would have supporting roles. The termination of women's aviation programs in the services after World

Fig. 4.2 Jerrie Cobb, one of the 'Mercury 13' as described in the text who passed astronaut screening successfully and then were denied access to the astronaut program. Top: Jerrie standing by Mercury capsule. Bottom: Jerrie in NASA Multi-Axis Space Test Inertia Facility (MASTIF). (Courtesy of NASA)

War II meant that no women had any access to current (jet) technology. Yet they passed the strict physical and psychological tests, some of them doing as well or better than the male candidates. The program was cancelled due to lack of funds for its continuance.

The second and third group of astronaut candidates (1962 and 1963), included Armstrong, Borman, Conrad, Lovell, McDivitt, See, Stafford, White, Young, Aldrin, Anders, Bassett, Bean, Cernan, Chaffee, Collins, Cunningham, Eisele, Freeman, Gordon, Schweickart, Scott, and Williams. Age restrictions were slightly lowered (35 for second group, 34 for the third group), educational qualifications broadened to include biological sciences, flight experience not limited to the military, and flight hours lowered to 1000 h for third group. A total of 23, including the 'new nine' from the second group and the 'Fourteen' from the third group, most of them military officers, were selected in these groups (Fig. 4.3).

In 1965, in response to the Space Science Board of the National Academy of Sciences recommendations in the report "Apollo Experiments and Training on the Scientific Aspects of the Apollo Program" [9], NASA diverged from its original focus on operational requirements only and included 'strong scientific training' as an additional criterion for astronaut selection, though emphasis was still on the former. Six astronauts (the Scientists), all civilians and all with PhDs or MDs, including Garriott, Gibson, Graveline, Kerwin, Michel, and Schmitt, were selected in this group tailored for the Apollo program (Fig. 4.4). Similar criteria to group three were used to select the fifth group of 19 astronauts (Brand, Bull, Carr, Duke, Engle, Evans, Givens, Haise, Irwin, Lind, Lousma, Mattingly, McCandless, Mitchell, Pogue, Roosa, Swigert, Weitz, Worden) in 1966 (Fig. 4.4). In 1967, a sixth group of eleven astronauts (the Excess Eleven) including Allen, Chapman, England, Henize, Holmquest, Lenoir, Llewellyn, Musgrave, O'Leary, Parker, and Thornton, all civilians and all requiring a year of training in jet pilot school, was selected according to criteria used for Group 4. In 1969 seven astronauts (Bobko, Crippen, Fullerton, Hartsfield, Overmyer, Peterson, Truly) were transferred to NASA from the cancelled DOD Manned Orbiting Laboratory Program.

61% of the 500 astronauts that have traveled into space, most of them orbiting the Earth, since the flight of Yuri Gagarin, have been Americans. 62 astronauts have flown as part of the Apollo or Apollo precursor (Mercury, Gemini) programs. What was similar and what was distinctively different about the Apollo astronauts? The astronaut selection and many aspects of the training process have been similar: intensive physical and medical screening, academic expertise in aerospace-related topics, spacecraft systems and safety training, team building exercises, roles as not only operators but integral

Fig. 4.3 Second and third group of astronauts: Top: New Nine selected for Gemini: L to R Back See, McDivitt, Lovell, White Stafford. L to R Front: Conrad, Borman, Armstrong, Young. Bottom: The Fourteen for Gemini and Apollo: L to R Back: Collins, Cunningham, Eisele, Freeman, Gordon, Scheickart, Scott, Williams. L to R Front Aldrin, Anders, Bassett, Bean, Cernan, Chafee. (Courtesy of NASA)

participants in the development and testing of equipment, and intensive simulation training. The distinctive character of the Apollo astronaut experience and contribution, including their unique role in developing the first successful extraterrestrial exploration program, will be discussed in the chapters that follow.

The Apollo astronauts were successful in doing what they were trained to do, including characterizing the lunar surface and obtaining representative samples from the Apollo landing sites. The average surface compositions

Fig. 4.4 Top: The Scientists (Group 4): L to R Front: Michel (physicist), Schmitt (astrogeologist), Kerwin (physician). L to R back: Garriott (Engineer), Gibson (Engineer). Graveline (physician) missing. Bottom: Original Nineteen (Group 5): L to R back: Swigert, Pogue, Evans, Weitz, Irwin, Carr, Roosa, Worden, Mattingly, Lousma. L to R Front Givens, Mitchell, Duke, Lind, Haise, Engle, Brand, Bull, McCandless. (Courtesy of NASA)

obtained from orbital geochemical instruments with larger fields of view encompassing the landing sites match such averages for the samples collected extremely well [10]. The impact of extensive and intensive training in focused, systematic approaches led by dedicated, experienced field geologists cannot be overestimated. This training involved hundreds of hours of simulated lunar surface traverses at terrestrial analog sites for the Apollo astronauts [10] as well as for the 'backroom', surface and orbital experiment and tool developers, control room crew, flight managers and engineers, with the idea in mind that everyone would have an understanding of the approach and a common language. This was not classical geological field camp. Their training involved highly selective sampling, with a limited number of samples and amount of time allowed. Samples were to represent the processes seen beforehand in aerial coverage of the surrounding area and then on the ground. As their gloves allowed limited manual dexterity, they were trained in oral documentation techniques [11]. The equipment and tools the astronauts used were developed, tested, and modified through early time motion studies. In simulations of lunar surface operations, astronauts systematically coordinated their documentation and sampling techniques until their activities became an efficient choreography, as clearly indicated in the lunar surface journals.

The online Apollo Surface Journal [12], a second-by-second voice transcription with annotated time-stamped links for all astronaut surface activity, including sampling and photography, is being utilized extensively in this book as an invaluable means of analyzing astronaut performance in Apollo surface activities. In the following chapters, actual performance parameters are considered including capability for distance on foot or in rovers, documentation and sampling of field stations, and manual operation of tools and instruments, all as a function of time. The application of these analyses as 'lessons learned' for planning the next generation of field science activities on the Moon and elsewhere are considered here as well (Fig. 4.5).

4.3 Astronaut Training Activities and Their Rationale

The field (and flyover) sessions provided essential training for the astronauts [13]. Through field trips to local sites and overflights of surrounding regions beforehand, the astronauts became skilled in use of maps for understanding regional context for interpretation of local geology. They learned most effective and efficient ways to 'sample', visually assess, and document structures that would allow interpretation of a local landscape in its regional context, critical when available (EVA) time and (sample) mass are limited resources. As the astronauts engaged in field site activities, they became proficient in the use of geology field tools provided and could provide feedback to the designers for

Fig. 4.5 Top: Excess 11 (Group 6): L to R front Chapman, Parker, Thornton, Llewellyn. L to R back Allen, Heinze, England, Holmquest, Musgrave, Lenoir, O'leary. Bottom: Quarantine box for Apollo 11 rocks being unloaded at Ellington Air Force Base near Receiving Lab. NASA 'Official Witnesses' George Low, Samuel Phillips, Thomas Paine, and Robert Gilruth pose with the box. (Courtesy of NASA)

the refinement of these tools. How did each of the training session contribute to the development of these capabilities?

Much of the training, especially the early training and later simulations, was done in the area around Flagstaff, Arizona, location of the USGS Astrogeology Branch, which provided much of the expertise and tools for astronaut field training. The surrounding area provided access to a great range of well documented volcanic features (calderas, cinder cones, flows), a documented impact feature (Meteor Crater), and an area of extensive physical erosion (Grand Canyon) with minimal vegetation cover, and thus excellent exposure of underlying stratigraphy. Starting in the third year of training (1965), the astronauts regularly engaged in simulation activities known as 'playing the Moon game' [14]. Pairs of astronauts were placed in field locations for which they had little prior knowledge, as they would when seeing lunar sites for the first time. They had to plan traverses to collect samples representing and highlighting major geological events that had occurred in the surrounding area. They communicated with instructors by radio, and those conversations were recorded, and later discussed and analyzed to improve the astronaut's interpretive ability and communication skills. Many sites were visited several times. See the Training Schedule on Tables 4.2, 4.3, 4.4, 4.5, 4.6, 4.7, and 4.8.

4.4 1963–1968 Astronaut Training

Some of these sites were visited several times. From mid-1965 on, the astronauts engaged in 'the moon game' when visiting these sites.

Cinder lake crater fields 1 and 2 [15] (1967–1968) (Figs. 4.6 and 4.7), AZ, Black Mesa Crater field, Verde Valley (1970), AZ (Apollo 11, Apollo 12, Apollo 13); Operation Snowball Crater (1964), Medicine Hat, Alberta. In Arizona, at NASA's request, the USGS used explosives to create craters in existing multi-episodic volcanic cinder fields, in attempts to make an analog of the lunar landscape and regolith, as understood at that time. Vehicles like the Explorer (Fig. 4.6) were used to get around these heavily cratered sites. In particular, an attempt was made to recreate the earliest landing site (Apollo 11, Mare Tranquillitatis). The USGS used available imagery of the lunar surface, including images from Lunar Orbiter, to mimic topography on a 1:1 scale for those sites, producing craters ranging from feet to tens of feet in size. Astronauts also studied the TNT-generated Operation Snowball crater in Medicine Hat, Alberta. The sites were used not only for training and testing of equipment on the ground, but also, during overflights, to test the

Table 4.2 Astronaut field training 1963 and 1964

Dates	Location	Astronauts
1963, January, 3 days	Meteor Crater and San Francisco Volcanic Field (volcanic features), observe Moon at Lowell Observatory, Flagstaff, AZ	Armstrong, Borman, Conrad, Lovell, McDivitt, See, Stafford, White, Young
1964, March 4–7, first half	Grand Canyon, Arizona	Carpenter, Shepard, Armstrong, See, Collins, Cunningham, Eisele, Freeman, Gordon, Schweickart, Scott, Williams.
1964, March 11–14, second half	Grand Canyon, Arizona	Aldrin, Anders, Bassett, Bean, Cernan, Chaffee, Schirra, Slayton, Grissom, Cooper, White, Borman, McDivitt, Lovell, Conrad, Young, Stafford
1964, April 2–3	Marathon Basin and Big Bend Park, Texas	Aldrin, Anders, Bassett, Bean, Borman, Cernan, Chaffee, Collins, Cunningham, Cooper, Eisele, Freeman, Gordon, Schweikart, Scott, Sheppard, C.C. Williams, Young.
1964, April 15–16	Marathon Basin and Big Bend Park, Texas	Carpenter, Schirra, Armstrong, Conrad, Lovell, McDivitt, See, Stafford, White
1964, April 30–May 3	Kitt Peak Observatory. Observe the Moon.	All astronauts at that time (First Seven, New Nine, The Fourteen)
1964, May 20–23	San Francisco Volcanic Field Flyovers	All astronauts at that time (First Seven, New Nine, The Fourteen)
1964, June 3–5	Philmont Boy Scout Ranch, Cimarron, NM; geological mapping, geophysical instrument traverses	Aldrin, Anders, Armstrong, Bassett, Cernan, Chaffee, Collins, Conrad, Cooper, Cunningham, Eisele, Freeman, Gordon, Lovell, Bean, Schweikart, Scott, See, White, C.C. Williams
1964 October 7–9	Newbury Crater, Bend, Oregon	Cernan, Chaffee, Collins, Schweikart, Scott
1964 October 15–17	Newbury Crater, Bend, Oregon	Aldrin, Armstrong, Bassett, Bean, Cunningham, See, C.C Williams
1964 October 23–25	Valles Caldera, New Mexico	Aldrin, Anders, Bean, Chaffee, Collins, Freeman, C.C. Williams
1964 October 29–31	Valles Caldera, New Mexico	Armstrong, Bassett, Cernan, Cunningham, Gordon, Scott, See
1964 November 13–14	Valles Caldera, New Mexico	Eisele, Schweikart

Table 4.3 Astronaut field training 1965

Dates	Locations	Astronauts
1965 January 12–14	Big Island Hawaii	Aldrin, Anders, Armstrong, Bassett, Bean, Chaffee, Collins, Conrad, Cunningham, Eisele, Gordon, Schweikart, Scott, See, C.C Williams
1965 February 17–18	Nevada Test Site, Study nuclear crater	Aldrin, Armstrong, Bassett, Bean, Cernan, Collins, Carpenter, Collins, Cunningham, Eisele, Freeman, Gordon, Schweickart, Scott, Williams, Aldrin, Anders, Bassett, Bean, Cernan, Chaffee.
1965 February 24–25	Nevada Test Site, Study nuclear crater	Cunningham, Eisele, Gordon, Schweikart, Scott, Schirra, Slayton, Grissom, Cooper, White, Borman, McDivitt, Lovell, Conrad, Young, Stafford
1965 March 3–4	Nevada Test Site, Study nuclear crater	Chaffee, See
1965 April 22–23	Meteor Crater, Winslow, AZ	Anders, Bassett, Cernan, Chaffee, Collins, Cunningham, Eisele, Gordon, Schweikart, Scott, C.C Williams
1965 June 29–July 2	Katmai, Alaska	Aldrin, Anders, Bassett, Cernan, Chaffee, Cunningham, Schweikart, Scott, C.C. Williams
1965 July 12–16	Iceland (volcanics)	Anders, Bassett, Cernan, Chaffee, Cunningham, Eisele, Schweikart, Scott, C.C. Williams
1965 September 1–3	Medicine Lake, CA	Anders, Bean, Schweikart
1965 September 8–10	Medicine Lake, CA	Bassett, Cunningham, C.C. Williams
1965 September 14–21	Zuni Salt Lake, NM	Anders, Bean, Cunningham, Schweikart
1965 November 8–10	Pinacate Volcanic Field, northern Mexico	Anders, Cunningham, Schweikart, C.C Williams
1965 December 27–29	Zuni Salt Lake, NM	Bassett, Cernan, Chaffee

astronaut's ability to know their location on the map during overflights and on the ground.

Nevada Test Site, nuclear detonation craters (Fig. 4.8): The astronauts performed numerous geological and geophysical studies, collecting and characterizing impact related features on samples collected at several of these craters, including Sedan, Schooner, and Buckboard Mesa craters. They also explored several ancient volcanic formations at nearby Timber Mountain caldera.

Table 4.4 Astronaut field training 1966, 1967, and 1968

Dates	Locations	Astronauts
1966 June 2–3	Grand Canyon, AZ	Brand, Bull, Carr, Duke, Engle, Haise, Irwin, Lind, Lousma, Mattingly, Mitchell, Pogue, Roosa, Swigert, Weitz, Worden
1966 June 23–24	West Texas, marathon Basin, Big Bend	Brand, Bull, Duke, Engle, Evans, Givens, Haise, Irwin, Kerwin, Lind, Lousma, Mattingly, McCandless, Michel, Mitchell, Pogue, Roosa, Schmitt, Swigert, Weitz, Worden
1966 July 27–29	Newberry crater, Bend, Oregon	Brand, Bull, Duke, Engle, Evans, Givens, Haise, Irwin, Kerwin, Lind, Lousma, Mattingly, McCandless, Michel, Mitchell, Pogue, Roosa, Schmitt, Swigert, Weitz, Worden
1966 August 21–25	Katmai, Alaska, valley of ten thousand smokes	Brand, Bull, Carr, Duke, Engle, Evans, Garriott, Gibson, Givens, Haise, Irwin, Kerwin, Lind, Lousma, Mattingly, McCandless, Michel, Mitchell, Pogue, Roosa, Schmitt, Swigert, Weitz, Worden
1966 September 25	Valles Caldera, NM	Brand, Bull, Carr, Duke, Engle, Evans, Garriott, Gibson, Givens, Haise, Irwin, Kerwin, Lind, Lousma, Mattingly, McCandless, Michel, Mitchell, Pogue, Roosa, Schmitt, Swigert, Weitz, Worden
1966 November 7–10	French and Hopi Buttes, AZ	Schmitt, Engle, Weitz
1966 November 29–December 2	Pinacate Volcanic Field, northern Mexico	Brand, Bull, Carr, Duke, Evans, Gibson, Haise, Irwin, Lind, Lousma, Mattingly, Michel, Mitchell, Pogue, Schmitt, Swigert, Weitz, Worden
1967 February 12–19	Big Island Hawaii	Bull, Carr, Duke, Engle, Pogue, Garriott, Gibson, Haise, Irwin, Lousma, Mattingly, McCandless, Michel, Mitchell, Worden
1967 March 16–17	Pinacate Volcanic Field, northern Mexico	Engle, Garriott, Kerwin, McCandless
1967 March 20–24	Big Island Hawaii	Bull, Carr, Duke, Engle, Pogue, Garriott, Gibson, Haise, Irwin, Lousma, Mattingly, McCandless, Michel, Mitchell Worden
1967 May 16–19	Zuni Salt Lake, NM; Hopi Buttes, Meteor Crater, AZ	Brand, Carr, Duke, Evans, Garriott, Haise, Irwin, Kerwin, Lind, Lousma, Mattingly, McCandless, Michel, Mitchell, Roosa, Swigert, Weitz, Worden
1967 May 31–June 2	Zuni Salt Lake, NM	Bull, Engle, Givens, Pogue
1967 July 2–8	Iceland (volcanics)	Anders, Armstrong, Brand, Carr, Duke, Engle, Evans, Garriott, Gibson, Haise, Lind, Lousma, Mattingly, Michel, Mitchell, Pogue, Schmitt, Swigert, Worden
1968 February 13–14	Newberry Crater, Bend, Oregon	Borman, Aldrin, Lovell, Armstrong, Collins, Anders

Table 4.5 Astronaut field training, assigned Apollo 11–14 prime and backup crews 1969

Dates	Locations	Astronauts
1969 February 24	Apollo 11 training, **final field simulation**, Sierra Blanca TX	Armstrong, Aldrin, Lovell, Haise
1969 March 1–14	Apollo 12 training, Quitman Mountains	Conrad, Bean
1969 April 8–9	Apollo 12 training, Kilbourne Hole, NM	Conrad, Bean, Gibson
1969 May 1–2	Apollo 12 training, Big Bend, TX	Conrad, Bean, Gibson
1969 July 10	Apollo 12 training, Meteor Crater, AZ	Bean
1969 August 9–11	Apollo 12 training, Big Island Hawaii	Conrad, Gibson, Bean, Scott, Irwin, Schmitt
1969 October 10	Apollo 12 training, **final field simulation**, Sunset Crater, AZ	Conrad, Bean, Scott, Irwin, Gibson
1969, mid to late September	Apollo 13 training, Orocopia Mountains, CA	Lovell, Haise, Young, Duke, Schmitt
1969 September 24–October 1	Apollo 13 training, Mono Crater, CA	Lovell, Haise, Young, Duke
1969 24 October	Apollo 13 training, Meteor Crater, AZ	Lovell, Haise, Young, Duke
1969 November 11	Apollo 13 training, Kilbourne Hole, NM	Lovell, Haise, Young, Duke
1969 December 17–20	Apollo 13 training, Big Island Hawaii	
1969 August 14	Apollo 14 training, East Flagstaff Volcanic Fields	Shepard, Mitchell, Engle
1969 August 22–23	Apollo 14 training, Craters of the Moon, ID	Shepard, Mitchell, Cernan, Engle

Table 4.6 Astronaut field training, assigned Apollo 13–16 prime and backup crews 1970

Dates	Locations	Astronauts
1970 March 15–16	Apollo 13 training, **Final field Simulation,** USGS Black Canyon Crater Field, Verde Valley, AZ	Lovell, Haise, Young, Duke
1970 February 14–18	Apollo 14 training, Pinacate volcanic fields, northern Mexico	Shepard, Mitchell, Cernan, Engle
1970 April 2–4	Apollo 14 training, Big Island Hawaii	Shepard, Mitchell, Cernan, Engle
1970 June 3–4	Apollo 14 training, Kilbourne Hole, NM	Shepard, Mitchell, Cernan, Engle
1970 June 18–19	Apollo 14 training, east Flagstaff volcanic fields	Roosa, Evans
1970 August 11–13	Apollo 14 training, Ries Crater, Germany	Shepard, Mitchell, Cernan, Engle
1970 September 11	Apollo 14 training, Nevada Test Site, nuclear crater	Shepard, Mitchell, Cernan, Engle, Evans

Table 4.6 (continued)

Dates	Locations	Astronauts
1970 November 16	Apollo 14 training, **final field simulation**, Black Mesa crater field, Verde Valley, AZ	Shepard, Mitchell
1970 May 6–8	Apollo 15 training, Chocolate Mountains, CA	Scott, Irwin, Gordon, Schmitt
1970 May 11–12	Apollo 15 training, Chocolate Mountains, CA	Scott, Irwin, Gordon, Schmitt
1970 June 3	Apollo 15 training, Orocopia Mountains, CA	Scott, Irwin, Gordon, Schmitt
1970 June 4–5	Apollo 15 training, East Flagstaff Volcanic Field, AZ	Scott, Irwin, Gordon, Schmitt
1970 June 15–17	Apollo 15 training, Merriam Crater, Flagstaff, AZ	Scott, Irwin, Worden, Gordon, Schmitt, Brand
1970 July 8–10	Apollo 16 training, San Juan Mountains, NM??	Young, Duke, Haise, Pogue, Carr
1970 July 22–23	Apollo 15 and 16 training, Eagle Butte Crater, Medicine Hat, Alberta, Canada	Gordon, Brand, Schmitt, Scott, Irwin, Worden, Young, Duke, Haise, Pogue, Carr
1970 August??	Apollo 15 training, Cinder Lake Crater Field, Flagstaff, AZ	Scott, Irwin, Worden, Gordon, Schmitt, Brand, Allen, Heinz, Parker
1970 August 26–28	Apollo 15 training, San Juan Mountains, CO	Scott, Irwin, Gordon, Schmitt
1970 September 1–2	Apollo 16 training, Colorado plateau	Young, Duke, Haise, Pogue, Carr
1970 September 17–18	Apollo 15 training, Buell Park, AZ	Scott, Irwin, Gordon, Schmitt
1970 October 7–9	Apollo 15 training, northern Minnesota	Scott, Irwin, Gordon, Schmitt
1970 October 12–14	Apollo 16 training, northern Minnesota	Young, Duke, Haise, Pogue, Carr
1970 November 2–3	Apollo 15 training, Merriam Crater and Cinder Lake Crater Field, first test Grover LRV simulator, Flagstaff, AZ	Scott, Irwin, Schmitt, Cernan
1970 November 12–13	Apollo 16 training, Nevada test site, nuclear crater	Young, Duke
1970 November 23–24	Apollo 16 training, San Gabriel Mountains, CA	Young, Duke
1970 December 5–12	Apollo 15 training, Big Island Hawaii	Scott, Irwin, Gordon, Parker, Schmitt, Allen

Table 4.7 Astronaut field training, assigned Apollo 15–17 prime and backup crews, 1971

Dates	Locations	Astronauts
1971 January 18	Apollo 15 training, Kilbourne Hole, NM	Scott, Irwin, Gordon, Schmitt
1971 January 18–20	Apollo 16 training, Kilbourne Hole, NM	Young, Duke, Haise
1971 February 10–12	Apollo 15 training, Ubehebe Craters, CA	Scott, Irwin, Gordon, Schmitt
1971 February 25–26	Apollo 16 training, East Flagstaff volcanic fields, AZ	Young, Duke, Haise
1971 March 11–12	Apollo 15 training, Rio Grande Canyon, Taos, NM	Scott, Irwin, Gordon, Schmitt
1971 March 29–30	Apollo 16 training, Merriam Crater, Flagstaff, AZ	Young, Duke, Haise
1971 April 26–27	Apollo 16 training, Black Canyon Crater Field Camp, Camp Verde, AZ	Young, Duke, Haise
1971 April 29–30	Apollo 15 training, Coso Hills, CA	Scott, Irwin, Worden, Gordon, Schmitt, Brand
1971 May 5–6	Apollo 16 flyover training, San Francisco Peaks volcanic fields	Mattingly, Roosa
1971 May 20–21	Apollo 15 training, Nevada test site, nuclear crater	Scott, Irwin, Gordon, Schmitt
1971 May 24–25	Apollo 16 training, Capulin Mountains, NM	Young, Duke, Haise
1971 June 10–11	Apollo 16 training, Mono Crater, CA	Young, Duke, Haise
1971 June 23–25	Apollo 16 flyover training, Craters of the Moon, Butte Crater (rille), landslides and mud flows (in Idaho); Pleistocene lake terraces, Salt Lake, structure, Wasatch Mountains (in Utah)	Mattingly, Roosa
1971 June 25	Apollo 15 training, **Final field Simulation**, Coconino point, near Flagstaff, AZ	Scott, Irwin, Gordon, Schmitt
1971 June 29–30	Apollo 16 training, Merriam Crater, Cinder Lake Crater Field, Flagstaff, AZ	Young, Duke
1971 July 7–9	Apollo 16 training, Sudbury, Ontario, Canada	Young, Duke, Haise
1971 July 19–21	Apollo 16 flyover and field training; Pinacate Volcanic Field (Mexico and New Mexico; Meteor Crater, Hopi Buttes, Arizona	Mattingly, Roosa
1971 September 7–8	Apollo 16 flyover training, Rio Grande Valley, and Valles Caldera, New Mexico	Mattingly, Roosa
1971 September 9–10	Apollo 16 training, Rio Grande Canyon, Taos, NM	Young, Duke, Haise, Mitchell
1971 October 13–15	Apollo 16 flyover training, Mt., Lassen, Medicine Lake Highlands, Crater Lake Caldera, CA and OR	Mattingly, Roosa
1971 October 19–21	Apollo 17 training, Big Bend, TX	Cernan, Schmitt
1971 October 27	Apollo 16 training, Nevada test site, nuclear crater	Young, Duke, Haise, Mattingly

4 The Preparation: Planning, Selecting, Training   117

Table 4.7 (continued)

Dates	Locations	Astronauts
1971 October 27–28	Apollo 16 flyover training, Nevada test site, nuclear crater	Mattingly, Roosa
1971 November 17–18	Apollo 16 and 17 training, Coso Hills, CA	Young, Duke, Haise, Mattingly, Cernan, Schmitt
1971 November 17–18	Apollo 17 training, Merriam Crater, Cinder Lake Crater Field, Flagstaff, AZ	Evans
1971 December 7–13	Apollo 16 training, Big Island Hawaii	Young, Duke, Haise, Mattingly
1971 December 20–21	Apollo 17 training, Kilbourne Hole, NM	Cernan, Schmitt

Table 4.8 Astronaut field training, assigned Apollo 16 and 17 prime and backup crews, 1972

Dates	Locations	Astronauts
1972 January 24–25	Apollo 17 training, McCullough Mountains, Boulder City, NV	Cernan, Schmitt, Evans
1972 February 10	Apollo 16 training, **Final Field Simulation EVA 1, KSC**	Young, Duke
1972 February 17–18	Apollo 16 training, **EVA 2 and 3 final field simulation**, McCullough Mountains, Boulder City, NV	Young, Duke, Haise, Mitchell
1972 February 22–25	Apollo 17 training, Chocolate Mountains, Mojave Desert, CA	Cernan, Schmitt, Scott, Irwin
1972 February 23–24	Apollo 17 training, Merriam Crater, Cinder Lake Crater Field, Flagstaff, AZ	Evans
1972 March 14–15	Apollo 17 training, Sierra Madera	Cernan, Schmitt
1972 March 16–17	Apollo 16 training, Nevada test site, nuclear crater	Young, Duke
1972 April 10–12	Apollo 17 training, San Gabriel Mountains, CA	Cernan, Schmitt, Scott
1972 May 24–25	Apollo 17 training, Sudbury Crater, Ontario, Canada	Cernan, Schmitt, Evans
1972 June 22–29	Apollo 17 training, Big Island Hawaii	Cernan, Schmitt, Evans
1972 July 24–25	Apollo 17 training, Stillwater, Montana	Cernan, Schmitt, Evans, Young
1972 August 7–8	Apollo 17 training, Nevada test site, nuclear crater	Cernan, Schmitt, Young, Duke, Roosa
1972 September 6–7	Apollo 17 training, Tonopah, NV lunar crater	Cernan, Schmitt, Young, Duke
1972 October 6	Apollo 17 training, Blackhawk slide, Mojave Desert	Cernan, Schmitt, Young, Duke
1972 November 1	Apollo 17 training, **Final field simulation EVA 1**, sand pile, KSC	
1972 November 2–3	Apollo 17 training, **Final field simulation EVAs 2 and 3**, Sunset Crater, Cinder Lake Crater Field, Flagstaff, AZ	Cernan, Schmitt, Young, Duke

Fig. 4.6 USGS 'Explorer' vehicle used to transport astronauts around cratered sites. (Courtesy of USGS)

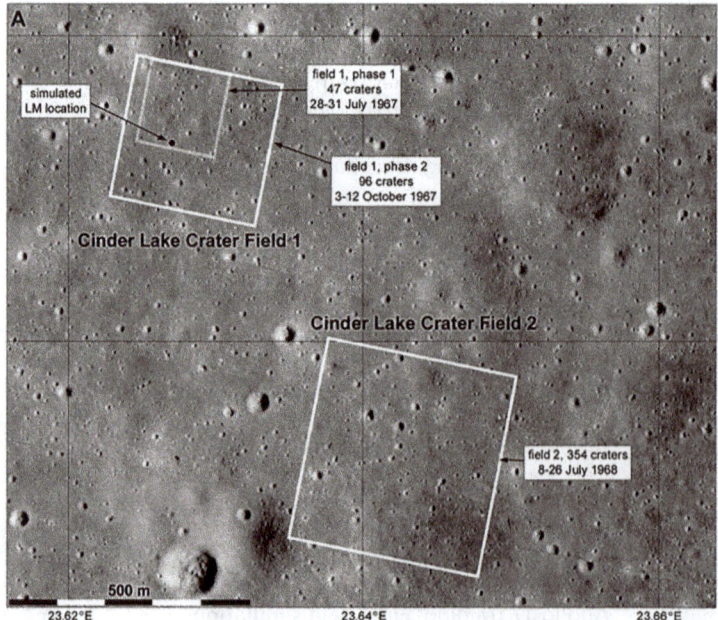

Fig. 4.7 Cinder Lake Crater Fields created to simulate the lunar surface. Note simulated LM location in Crater Field 1. (Courtesy of Phil Stooke, Personal Communication)

Fig. 4.8 Cernan and Schmitt at Nevada test site. (Courtesy of NASA)

Meteor Crater, Ries Crater, Germany. (Apollo 12, Apollo 13) As one of a relatively small number of confirmed (at that time) natural terrestrial impact craters, and one that was recent (40,000 years), thus the best-preserved and near Flagstaff, Meteor Crater was selected for training astronauts in the identification, and collection of samples for the geological interpretation of the most ubiquitous lunar feature, an impact crater. Ries Crater was also visited for this purpose.

Grand Canyon (Fig. 4.9): Because of the excellent exposure of underlying stratigraphy and physical erosion, the Grand Canyon was used, especially early on, to apply geological principles the astronauts had been exposed to in a classroom setting, including how to find, visually identify and collect samples representing surrounding rock types. The astronauts were also trained in determining their location on topographic maps and orbital images along their routes on major trails.

San Francisco Peaks (caldera) and San Francisco (East Flagstaff)f volcanic fields, including Merriam and Sunset Cinder Cones (Figure); French and Hopi Butte Volcanic Fields (Hopi reservation, NE Arizona) (Apollo 12): The landscape around Flagstaff illustrates the range of structures associated with an area of extensive volcanic activity, anticipated at that time to be analogous to areas on the Moon in terms of regolith origin and structures. These features included a stratovolcano (San Francisco Peaks) its caldera, and surrounding smaller-scale volcanic cinder cones (Merriam and Sunset Craters), vents, and flows. As such, this area represented an excellent place for training and testing equipment and procedures.

Fig. 4.9 Unidentified trainer and astronaut at Grand Canyon. (Courtesy of NASA and USGS)

Philmont Boy Scout Ranch, Texas. The USGS had just completed a major geological survey for this area, documenting its complicated structure and associated distinctive major rock types, providing an opportunity for the astronauts to identify and select representative igneous and sedimentary rock types, practice orienteering with geological maps, and perform structural analysis from strike and dip measurements with a Brunton Compass.

Big Bend, Marathon Basin, Texas (Fig. 4.10) **(Apollo 12)**: This site represented an opportunity for the astronauts to study well exposed structural and stratigraphic relationships reflecting an amazing amount of geological diversity. Underlying rocks are a Paleozoic marine sedimentary basin, which experienced uplift and extensive folding when western and eastern pieces of North America collided on either side, creating the Ouachita mountains and forming the supercontinent Pangea. The range was subsequently eroded, exposing extremely resistant distinctive white chert seen today. Volcanic rocks are also exposed along the Rio Grande River canyons.

Fig. 4.10 Cernan at Big Bend. (Courtesy of NASA)

Fig. 4.11 Unidentified astronauts at Newberry Crater, Oregon with local expert Aaron Waters (holding box) discussing origins of volcanic features. (Courtesy of NASA)

Newberry Crater, Bend, Oregon (Fig. 4.11) **and Valles Caldera, Jemez Mountains, New Mexico**: Both of these sites represented large volcanic caldera complexes. At Newberry Crater, astronauts traversed and sampled complex volcanic structures, including nested craters, and rocks representing an extensive range of volcanic activities (obsidian flows, pumice cones, cinder cones, tuff (consolidated ash) rings on a large shield volcano. At the Newberry site, the astronauts were asked to gather evidence to support either an impact or volcanic origin for a crater of disputed origin, Hole-in-the-Ground. The then visited Valles Caldera to inspect a complex of strikingly different structures and rocks, including rhyolite domes in the center of the caldera, active springs and fumaroles, welded tuffs, pumice falls, and ash flows, and asked to discuss the differences between the two sites.

Big Island, Oahu, Hawaii (Fig. 4.12): At the world's largest volcano system, including the world's largest active volcano, Mauno Loa, the astronauts traversed the range of fresh volcanic features, including calderas, the summit crater of Mauna Loa, recent eruption of Kilauea Iki, lava flows, lakes, and gas and lava vents, including more eroded, extinct ones on Oahu, pit craters, and lava tubes. They were asked to plan trajectories which would allow them to sample units identified on geological maps. They utilized cuff checklists in the performance of assigned tasks at selected sites. In later traverses, rovers were used.

Katmai, Valley of Ten Thousand Smokes, Alaska (Fig. 4.13): This site is a rocky lunar-like volcanic landscape, with recent well documented eruptions and deep stream erosion exposing the history of volcanic deposits representing the type and duration of volcanic activities. The astronauts sampled these dust and ash deposits, and the extinct fumaroles for which the valley was named.

Askja Caldera, Iceland (Fig. 4.14): This site was considered to be the most lunar-like of all field areas, with volcanic structures and deposits that included calderas, ash cones, active volcanic vents, and a surface covered with cinders, pumice, and a variety of lava flows.

Medicine lake, Mt. Lassen, Crater Lake Caldera, CA; Zuni Salt Lake, New Mexico; Pinacates, Mexico (Fig. 4.15). All of these sites are volcanic terranes with extensive calderas, and the astronauts trained using 'the moon game' at all of these sites. The Medicine Lake Highland includes a five mile diameter caldera surrounded by a large area of volcanic flows and obsidian domes. Zuni salt lake is also a large volcanic crater, and has a structurally complex rim. The Pinacates are a large area of volcanic terrane with explosive craters and a ring of subsidence fractures. At both the Zuni Salt Lake and Pinacates sites the astronauts performed geophysical exercises, assisting in the laying out of traverses and learning how to use geophysical instruments in the field.

4 The Preparation: Planning, Selecting, Training

Fig. 4.12 Fred Haise and Jim Lovell at Big Island pahoehoe lava flow. (Courtesy of NASA)

Fig. 4.13 Two astronauts at Katmai Valley. (Courtesy of NASA)

Fig. 4.14 Astronauts Vance Brand and Stuart Roosa (left) with Iceland volcanics instructor (glasses) Ted Ross in Askja, Iceland. (Courtesy of NASA)

Fig. 4.15 Astronauts Ed Mitchell and Stuart Roosa learning to use a gravity meter at Pinacates volcanic region, Sonora, Mexico. (Courtesy of NASA)

4.5 1969–1972 Apollo Prime and Backup Crew Field Training

Note: Apollo crews also visited sites described in previous section, as noted.

Volcanic Features at Sierra Blanca, Quitman mountains, SE of El Paso (Figure); Kilbourne Hole, Organ Mountains, Desert Peaks, Texas (Fig. 4.16): (**Apollo 11, Apollo 12, Apollo 13**) Sierra Blanca is an older stratovolcano underlain by intrusive volcanic stocks and dikes with simultaneously formed exposed ash, breccia, and flows, selected as an analog for lunar maria terrain. The Apollo 11 prime and backup crews received their only training in the use of geology field equipment, including rock handling equipment, voice recorders and cameras, to select and identify rocks representing the processes that occurred here. Kilbourne Hole is the site of an underground

Fig. 4.16 Haise and Lovell (top), Armstrong and Aldrin (bottom) at Sierra Blanco. (Courtesy of NASA)

volcanic explosion, and was selected to familiarize the Apollo 11 crew with the kind structures (shallow lake or maar), and volcanic ejecta (tuffs (consolidated ash), xenoliths (preexisting rock fragments surrounded by solidified magma), breccias (welded fragments of preexisting rocks) which would result.

Orocopia Mountains (Colorado-Sonoran Desert), Chocolate Mountains (Mojave Desert), Little Chuckwalla Mountains (Mojave Desert), CA (Apollo 13, 14, 15, 16): Most of the Apollo crews came to the Orocopia Mountains east of the Coachella Valley to practice field work because this area, similar to Chocolate and Little Chuckwalla Mountains and relatively well known to their instructors, has a complex geology shaped by activity along the San Andreas Fault to the west, resulting frequent downslope movements, all well exposed due to mimimal vegetation.

Mono Craters, CA (Apollo 13, Apollo 16) (Fig. 4.17), **Ubehebe Craters, Death Valley, CA (Apollo 15); Sudbury Basin, Ontario, Canada (Apollo 16)** (Fig. 4.18): All of these sites were considered to be volcanic craters, Mono Craters and Sudbury Basin analogs for the Apollo 16 site, at the time. Mono Craters are a volcanic chain with craters, domes, and lava flows in eastern California. Sudbury Basin was also interpreted to be volcanic in origin. Ubehebe Craters are a series of young, overlapping broad low-relief volcanos formed by shallow steam explosions (maars). The astronauts took photographs and samples, and used them to interpret the geological histories of these sites.

Craters of the Moon (Apollo 14) (Fig. 4.19), **Butte Crater lava tubes, Wapi volcanic field, Ammon pumice quarries, Idaho**: A barren, lunar like landscape of lava flows, flow fronts, tubes, and pyroclastics, this site provided

Fig. 4.17 At Mono Craters site, Instructor Muelhberger (left) talks with Astronaut Young in the foreground. Behind them, Instructor Wones talks with Astronaut Duke. (Courtesy of NASA)

Fig. 4.18 Astronauts Duke and Young at Sudbury Basin. (Courtesy of NASA)

well defined, recent volcanic craters as well in contrast to the impact crater at meteor crater and thus an excellent analog for what were seen as volcanic sites the astronauts were scheduled to visit.

San Juan Basin, Monument Valley, San Juan Mountains, Buell Park kimberlite pipe, Colorado Plateau, AZ, UT, CO, NM (Apollo 15, Apollo 16). The Colorado Plateau is an uplifted region with underlying ancient volcanic structures, such as the Buell Park kimberlite, probably the largest one in the world, indicating a large volcanic explosion releasing ultramafic magma from the crust/mantle interface. The Plateau is extensively dissected with deeply carved canyons, buttes, mesas, and smaller plateaus, such as Monument Valley. The San Juan Basin is an asymmetrical depression in the southern portion of the Plateau, bordered on the east by the San Juan Mountains, formed of uplifted upturned strata of younger volcanic rocks in contact with Precambrian rocks of the San Juan Basin. The astronauts performed traverses in these complex and varied sites with well dissected and exposed stratigraphy to further develop their sample and site characterization skills and the use of geophysical tools and equipment.

Duluth, Minnesota; San Gabriel Mountains, CA; Stillwater Complex, Montana (Apollo 16, Apollo 17). All of these sites were selected because they have extensive exposures of anorthosites, and would give the astronauts training in identifying anorthosites and understanding their relationship to surrounding rocks. At Duluth anorthosites are intermixed with gabbros. The Stillwater is a complex structure, where igneous rocks, including ultramafics, gabbros, and anorthosites, intruded into metamorphic rocks. San Gabriel anorthosites, the oldest exposed rocks in California, formed from an intrusive igneous body.

Fig. 4.19 Unidentified Astronauts at Crater of the Moon, Idaho. (Courtesy of NASA)

Fig. 4.20 Astronaut Schmitt at the Rio Grande Gorge near Taos, NM. (Courtesy of NASA)

Rio Grande Gorge, Taos, NM (Apollo 15) (Fig. 4.20). This site was selected as an analog for Hadley Rille, both sites comparable in terms of depth and width and surrounded by basaltic lava flows. Unobstructed stretches along the gorge and rille allow rover traverses of comparable length.

Coso Hills, Coso Volcanic Field, CA (Apollo 15). This site is an analog for Apollo 15: low relief volcanic plains next adjacent to steeply sloping front of granitoid and metamorphic rocks, analogous to the Apennine hills.

Capulin Mountains, NM (Apollo 16). A cinder cone from which a series of volcanic flows emanate, this site provided an opportunity for the astronauts to distinguish and interpret volcanic flow sequences.

Wasatch Range, Great Salt Lake, Utah; McCullough Range, Nevada (Apollo 16). These sites, observed from a flyover, allowed the astronauts to observe the interface between extensive sedimentary deposits around the south end of the Great Salt Lake, remnants of its Pleistocene era precursor Lake Bonneville, and the extensively faulted Wasatch Range, as well as the McCullough Range, both part of the Basin and Range Province.

Sierra Madera Crater, TX (Apollo 17): This site, generally thought to be an impact crater due to the extensively studied shock features [16] provided an opportunity for the astronauts to study an analog of a lunar impact crater. The exercise utilized a backroom with Parker as CapCom and Irwin acted as one of the scientists. There was one walking and one riding traverse both of which started near the crater rim and proceeded towards the central peak.

Blackhawk Slide, Lucerne Valley, San Bernardino Mountains (Apollo 17) (Fig. 4.21). This site, with its prominent, large well-define landslide, was selected as an analog to features on the south massif in Taurus Littrow Valley, the Apollo 17 landing site. Figure credit NASA/METI/AIST/Japan Space Systems and US/Japan ASTER Science Team. From terra S/C.

Lunar Crater Volcanic Field, Tonopah, NV (Apollo 17) (Fig. 4.22). A large volcanic field, with features analogous to the East Flagstaff volcanic field,

Fig. 4.21 Black Hawk Landslide in foreground. (Courtesy of NASA/METI/AIST/Japan Space Systems and US/Japan ASTER Science. Team)

Fig. 4.22 Gene Cernan and Harrison Schmitt on EVA practice in Nevada Volcanic Field near Tonopah. (Courtesy of NASA)

providing further opportunity for astronauts to sample, characterize, and interpret sites similar to those anticipated for the Moon.

4.6 Field Simulations

Starting in 1969, the astronauts field trips frequently involved planning and simulating traverses, and using the geology tools they would be using on the Moon. During those times, the astronauts would be in radio contact with the trainers and geology team. The also received training in the setting up of the ALSEP packages. The field trips would typically be preceded by overflights or orbital map training. Each Apollo mission team had comprehensive final simulations, with mission control and the geology team 'backroom' in the loop via the capcom as they would during the actual mission, and application of time and sampling constraints.

Apollo 11 and Apollo 12: About 1 month before the Apollo 11 mission, the astronauts performed a simulated EVA traverse at Cinder Lake Crater Field. Craters had been generated to create an analog for their landing sites, approximating crater distributions at the Mare Tranquillitatis landing site. Similarly for Apollo 12, another area of the Cinder Lake Crater Field had been modified to create an analog of the Mare Serenitatis Landing Site, and the Apollo 12 team performed a simulated EVA traverse.

For the later Apollo missions, final field simulation included more extensive preplanning of traverses and activities in places with analogous geology, as indicated in the tables, with handcart for Apollo 14, and rover for the later missions. Following the preplanning, Apollo 13, 14 and 15 each had one full day of simulations for their geological traverses. Apollo 16 and 17 had one full day of final simulation for EVA 1 (ALSEP deployment), and 2 days for geology traverses of EVA 2 and EVA 3, all with mission control in the loop, as described above.

References

1. Schaber, G., Chronology of Activities from Conception through the End of Project Apollo 1960-1973, USGS Astrogeology Branch, Open File Report 2005-1190, http://pubs.usgs.gov/of/2005/1190/, 2005.
2. Bailey. N., G. Ulrich, and K. Edmonds, Apollo Applications Program Field Test 8 with Section on Task Analysis, USGS Technical Letter 26, 1967.
3. Regan, R., Preliminary Geophysical Report on Selected Geologic Test Sites, USGS Technical Letter Astrogeology 29, 1967.
4. Staff Writers, Historical Evolution of Astronaut Selection, Space Safety Magazine, August 7, 2016. https://www.spacesafetymagazine.com/space-exploration/astronaut-cosmonaut/historical-evolution-astronaut-selection/. visited 10/1/24
5. Dethloff Henry, Suddenly, tomorrow came…A History of the Johnson Space Center, NASA SP-4037, 1993.
6. Beyer, D.H. and S.B. Sells, Selection and Training of Personnel for Space Flight, 1-6, Flight Crew Operations Subseries, JSC History Office, 1957.
7. Klein, C., The Mercury 13: Met the woman astronauts grounded by NASA, History.com, August 30, 2018. https://www.history.com/news/right-stuff-wrong-gender-the-woman-astronauts-grounded-by-nasa-2. visited 10/1/24
8. Ackmann, M. The Mercury 13: The untold story of thirteen American woman and the dream of space flight. Random House, 2003.
9. Ad Hoc Working Group on Apollo Experiments and Training, Scientific aspects of the Apollo Program, Lunar and Planetary Institute, 1963. https://www.lpi.usra.edu/lunar/documents/SonettReport.pdf. visited 10/1/24

10. Clark, P.E., Revolution in Field Science: Apollo approach to inaccessible surface exploration. Earth Moon and Planets, 106, 133-157. 2010.
11. M'Gonigle, J., D. Schleicher, and I. Lucchitta, A proposed scheme for lunar geologic description, USGS Interagency report: Astrogeology 18, 1969.
12. Jones, Eric, Apollo Lunar Surface Journal, NASA, 1995-2018. https://nasa.gov/historyalsj/. visited 10/1/24.
13. Nalewicki, J., Before going to the Moon, Apollo 11 astronauts trained at these five sites. Smithsonian Magazine, July 17, 2019. https://www.smithsonianmag.com/travel/going-moon-apollo-11-astronauts-trained-these-five-sites-180972452/. visited 10/1/24
14. Phinney, William, Science Training History of the Apollo Astronauts, NASA SP-2015-626, 2015.
15. Stooke, P., International Atlas of Lunar Exploration, 1st Edition, Cambridge University Press, 2008.
16. Wilshire, H.G. and T. Offield, Geoogy of the Sierra Madera cryptoexplosion structure, Pecos County, Texas, USGS Professional Paper 599-H, USGS, 1972.

5

Field Work Approach

5.1 Field Work Preparation

The philosophical framework for field work was consistent with approach used for other aspects of Apollo program activities, characterized by maximum flexibility and minimal micromanagement of multi-disciplinary teams, with varying degrees of experience [1]. These teams included astronauts, field geologists, and representative of those required for ground support with expertise in a variety of disciplines. Despite the critics of the 'Apollo Approach' as involving 'too many smart people spoiling the team' [2], this approach proved effective for practical problem solving and building esprit de corps. Enabling the seemingly impossible feat to bring human beings from the Moon and back with minimum loss of life in less than a decade required not only unprecedented pace for developing dramatically new technology, with management actively removing roadblocks rather than micromanaging. The unusual 'team-oriented' discipline and willingness to 'leave your ego at the door' to make this feat possible, not observed in Belbin's attempt [2] to create and study the behavior of Apollo-style management teams, was a result of the total commitment to the very risky 'dream' of space exploration and willingness to take risks, largely absent today, to assure America's role in it. This was an example set by Werner von Braun himself.

A series of documentary films were made by the USGS at field sites to describe the modification of terrestrial field and data analysis techniques for use by the Apollo program, to record the training of the astronauts in these techniques, and to record tests of supporting equipment. Many of these films, summarized in Tables 5.1, 5.2, 5.3, 5.4, 5.5, 5.6, 5.7, and 5.8 below, were

Table 5.1 Apollo Documentary Film Table 1

Title	Description
Apollo 16 lunar tool study	Evaluation of field tool construction and utilization
Grover geologic rover	Test and evaluate simulated lunar rover
Apollo 11 Lunar Rock Lab	Observe Lunar Surface from Apollo 11 orbit
Surveyor III Sunset Sequence	Observe Lunar Sunset from Surveyor III
Meteor Crater Field Trip NSF	Astrogeology Conference 1967. Astronaut training in aerial observation. Flagstaff to East Flagstaff volcanic field to meteor crater. Shoemaker impact geology training

Table 5.2 Apollo Documentary Film Table 2

Title	Description
Explorer	Performance of cargo carrier-like lunar rover for terrestrial site astronaut training
Rocket Flight	Test and Evaluation of astronaut rocket belt
Early Apollo Investigations Test 8	Hopi Buttes, AZ. May 1966. Test field techniques and tools (including camera) at complex geology site under simulated mission conditions for astronauts and ground crew. Entire process including pre-traverse study of remote photographs. Astronauts refer to same maps as ground crew
Apollo Applications Program Test 8	Hopi Buttes, AZ. May 1966. Testing navigation system for rovers J missions which included directional gyro, odometers, hybrid analog-digital signal processor with manual controls and readouts
Apollo Applications Program Test 6	Hopi Buttes, AZ. May 1966. Comparison traditional field methods, traversing the site, focusing on major features, recording observations on aerial photos, writing observations and recording pictures and samples in field notebook, to modified methods involving audio recording, transmittal and team analysis in real time, with tracking by relaying coordinates and plotting on the same grid and photo as used in field and using a mechanical plotter to record route. Also tried transmitting on laser system mounted on surveyor pole. Avoiding need to take notes… all transmitted orally. All data (orientation of camera, e.g.) to know what looking at. Rehearsal for 'back room'. Team looks at pictures for further analysis, and that is recorded too. Using calibrated overlay on screen, can measure grain sizes. Field instrumental analysis (XRD, XRS) also transmitted to team. Also tried active seismic experiment

archived and are now available online including "when the astronaut speaks" described in the last chapter. The films were recently moved to a USGS astroshare website and can now be found there in alphabetical order [3].

5 Field Work Approach

Table 5.3 Apollo Documentary Film Table 3

Title	Description
Ranger IX TV pictures of the Moon	Movie of all Ranger 9 pictures to impact
Ranger VIII TV pictures of the Moon	Movie of all ranger 8 pictures to impact
First Scientific Travers of mobile geological laboratory	Test of mobile laboratory not used on moon. Test using GRS under mobile field conditions for radioactive elements (no source)
Kilauea Volcano: geophysical studies and structural relations	Astronaut training. Volcanic structural geology. Seismograph and tiltmeters considered most useful. Structural (eruptions, calderas, fissures, collapses, faults and scarps concentric, slumping) flank type eruptions, earthquake epicenters (where magma generated), make cross section from surface data
Apollo Field Operations Test 3	Meteor Crater, AZ, May 18–20, 1965. Surveying staff, surveillance tv system. One man astronaut on traverse (surveying, stereo-periscope and surveying pole). Stereoscopic periscope in LEM plotted astronaut in field. Second man in LEM, receives data, keeps track of movement of astronaut, also field observer who was recording and timing operations for time motion studies and better tool design. Transmits to earth. Need to know orientation and context for photos and samples. Verbally describe what doing as well

Table 5.4 Apollo Documentary Film Table 4

Title	Description
Apollo Extension Systems Test 1	Hopi Butte, AZ. September 20–October 1, 1965. Area thought to be apollo analog. Much larger vehicle, which carried more cargo, than lunar rover. Sampling went with verbal description. Testing prototype tools and instruments. A gravimeter that worked! Penetrometer (in wheel tracks). Soil samples sieved in field into separate fractions. TV surveillance camera operated remotely. TV camera could look at local bedding in way in field folks couldn't. on foot recorded with tape recorder and timed. Also Gamma ray scintillometer to measure radioactivity. Big vehicle for deploying geophones along line. Drilled holes, plant charges for active seismography. Meter to determine magnetic susceptibility. XRD on samples. Ground in field. Microscopic analysis. Not in spacesuits. Inputs for equipment design. Further evolve to wearing spacesuits
When the astronaut speaks	See last chapter
Traverse 4, Part 2, Production 6710-21	In space suits with backpacks, carrying tool carrier. Using sampling equipment, surveying stick. Artificial crater site. No sound
Lunar Orbiter Production 2346. Boeing	Lunar Orbiter program. Main job to photograph potential landing sites
Surveyor test site Bonito Flow	Test remote control rover with balloon wheels on AA. Not doing well

Table 5.5 Apollo Documentary Film Table 5

Title	Description
Surveyor lunar roving vehicle film report 9	Testing three articulated axles in lab at small scale. Successful. Made another at full scale...also crossed barrier 2× wheel size. Tried on sand dune. Dacron covered spring wire wheels, wires through fabric acting as lug. Now pumice blacks. Works for soft or rugged surface. Surveyors stadia rod on vehicle making it possible to track vehicle and determine its distance. Worked on 30° sand slope. This was to be on surveyor? Also performed thermal test
Surveyor lander mirror test bonito flow	Built track across aa lava, bringing equipment. Little surveyor model in aa field. Testing model in field, transmitting data to station built adjacent to it
Modern suit test	In simulated space suit using simulated geological tools. Clumsy even without the gloves
Lawrence radiation lab technical film report	Project Sedan silent film. Cratering effects of high yield nuclear explosions in-ground
Mare Tranquillitatis	Flagstaff, AZ. Build model of m Tranquillitatis. In east flag cinder field: layers of cinders and clay. Drag ground (smooth surface). Mark positions of craters. Place explosives in holes so that get same size and age crater. Then do 'stratigraphy' of holes. Exploded in groups by age. See simulated M Tranquillitatis, see rays, secondaries

Table 5.6 Apollo Documentary Film Table 6

Title	Description
Early Apollo investigation test 13	Same as early apollo test 13 traverse 2 part II
Apollo applications program field test 3	Flagstaff, AZ. Feasibility studies for remote handling of thin sections. Filmed under microscope. Cross-polarized and not polarized. Assuming this equipment is on Moon. Later missions?
Apollo applications program	Later apollo missions, with roving vehicle. Navigational system test. Similar to Apollo Applications Program Test no. 8

Table 5.7 Apollo Documentary Film Table 7

Title	Description
Katmai Astronaut field trip	Mt Katmai Area, Alaska. Steaming volcanoes, recent ash flows. Aleutian range. Valley of 10,000 smokes. Intrusive granites also. Overlain by volcanic rocks. Glacial deposits along drainage. Mt. Katmai crater lake. Other volcanos erupted dacite. Trident flow, pyroclastics, pumice. Mostly ash. Loosely consolidated except where thickest. Rocks modified at vents. Fumaroles along fractures (steam and ash flows). Best example of flows. Still active in 1915
Hypervelocity impact and crater formation in rock and sand.	Simulate high velocity with block at different impact angles
Hawaiian volcanoes astronaut field trip	Finest examples shield volcano in world, cinder cones, flows, pyroclastic. Watch all kinds of flows. History of eruptions
Geology of Zuni Salt Lake Maar	Crater Kaje, short streams along steep sides flow into lake, no outlet. 200 ft deep. Dark volcanic cinder cones in center. Wall like sections of volcanic rock lie just below rim crest. Shallow, flat floored Volcanic explosion crater with walls of country rock and little lava. Lake used for salt. Maar like depressions on lunar surface? Training. Some older volcanic materials on slopes. Discontinuous wall like ridges around lake are dikes. Tuff and ash lies uncomfortably on underlying sedimentary beds. Compression by overlying layers then tensional features from rising domes. Saline lakes in centers of some surrounding small cones. Ground water warmed by residual heat in buried igneous mass. Springs carry soil dissolved from underlying sedimentary rocks. Gravity and magnetic anomalies
First Apollo scientific mission simulation	In early suit. Using survey stick. Focus on geological exploration. Small number of man hours available… necessary to produce as much data as possible in small amount of time. 1964 studies of what astronaut could do in space suit. Bonito flow on sunset crater. Gemini suit. Fake backpack. Nearby van served as lunar excursion module w/ surveillance TV system. Vidicon on surveying staff. Data to data reception center. Photogrammetric data. Vidicon, sun compass, clinometer (slopes), film camera for hi res photo. Sampling device near base (w/out stooping), penetrometer (mechanical properties) near base. Carried geophysical instrument package (incl seismometer) verbal description constant. Watched on TV. Describe regional setting to provide context of site. Critical need for precise descriptions. Humans most efficient system when properly trained

Table 5.8 Apollo Documentary Film Table 8

Title	Description
Early Apollo video tape field test rolls 1B	No sound. Poor quality snapshots
Early Apollo Video tape field roll 1B	Astronauts practice describing scene to 'mission ops'. Indicating where camera pointing, a panorama. Then go back to specific locations. Describing nature and relationships between stratigraphic units. When head off, describe soil and rocks along path and then more detail for destination along that path. Indicate when arrive at identified (from photomaps) unit. Interaction between Mission ops = backroom = local van with receiving equipment. Indicate color, albedo, texture, orientation, size (from soil particles to rocks). Use objects to indicate known scale to try to correlate soils and rocks, or observations from other stations
Early Apollo video tap field test rolls	No sound. Poor quality snapshots
Ealy Apollo test 13 traverse 2 Part II	Do sampling with Apollo material. Rays and secondaries. In manmade crater
Early Apollo test 13 traverse 4 part I	No sound. Astronauts in simulated suits with backpacks. Using survey stuck, deploying equipment. One astronaut taking pictures of this using carrier
Early Apollo field test roll 1A	Like early apollo video tape field roll 1-b. following man (not in spacesuits this time) around. In same way.... Albedos, colors, sizes small to large fragments in rocks and on ground, relationships. Context surroundings. Relationships between sites in terms of similar rocks or deposits. Minimum jargon. Refer to photomap units connected to the site, including what looks like on ground and on map. Describes contacts, positions in bedding, deposits

5.2 Mobility on the Lunar Surface

Clearly, the Apollo mission planning and simulations were based on the use of suited astronauts in a wheeled vehicle. At the time, driving was considered the most practical and least risky solution for the enhanced mobility required for the J missions. The use of space suits, the portable life support unit developed for walking, would allow the development of a wheeled device to enhance access without adding the cost and schedule that would be required for another life support system.

An alternative approach considered [4] was a jetpack, or Lunar Flying Unit (Fig. 5.1). Bell Aerosystems and North American Rockwell, who had already designed the Command/Service Module and the Lunar Module ascent rocket,

Fig. 5.1 Gene Shoemaker dons rocket belt during field training. (Courtesy USGS)

competed to develop the flying platform. Bell's platform allowed the astronauts to stand upright holding tools and samples and had an attachable platform for carrying another astronaut. Two rocket nozzles, one on either side, were controlled LM-like hand controls, and were capable of flying the astronaut up to 24 km 75 ft. above the surface. North American design was a flying chair with two sets of two rocket engines, for redundancy. The platform could fly the astronaut up to 8 km 2000 ft. above the ground. The never before used approach was considered potentially unstable and too risky, so NASA selected what became known as the 'moon buggy'. four all-wheel drive, battery powered 250 kg rover built by Boeing, designed, developed, and tested by General Motors Delco Electronics Division.

The Lunar Roving Vehicle (LRV) (Fig. 5.2) [5–8] was a 2.3-m wheel base, battery operated vehicle with each wheel independently controlled was created with an emphasis on simplicity in design and operation. The wheels, designed to prevent sinking in soft soil, were woven mesh made from zinc coated piano wire with titanium threads riveted in a chevron pattern. The webbing seats with Velcro strips interacted with the Velcro strips on the astronauts' life support backpacks, helping to keep them in place. The T-shaped hand controller allowed operation of the vehicle, including steering, speed, and braking, with one hand by one crew member. Testing of the LRV was done on springs to simulate lunar gravity. The LRV was completed in 17 months from contract go ahead to delivery!

To save space, the vehicle was folded up against the Lunar Module in a 4-foot cubic space for transit, and then deployed by the astronauts with the help of spring to release the vehicle and lock the wheels into place upon arrival on the surface. The simulation of the clever 'unfolding' can been seen in a NASA-produced video [9].

Basic requirements included the ability to cross craters 70 cm in diameter and clear rocks 75 cm in height. Chemical batteries allowed up to 100 km of travel. The onboard navigation system determined where the rover was in relationship to the lunar module at all times and allowed return the lunar

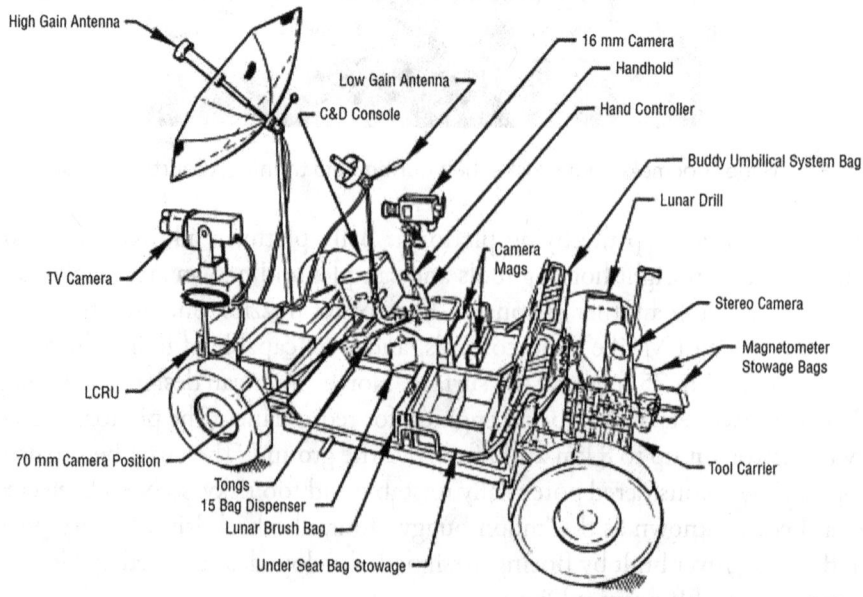

Fig. 5.2 The main components of the Lunar Roving Vehicle. (Courtesy of NASA)

module along the most direct route. The 2.3 m 210 kg vehicle could carry more than twice its weight (440 kg) which was to include two astronauts, lunar samples (up to 100 kg), RF equipment and a movie camera for real time communication with the Earth, cameras and the geological toolkit described below. Designed for speeds of up to 10 km/h, it actually achieved speeds approaching 20 km/h on later missions. The 'moon buggy' had sufficient traction to climb 25° slopes without significant wheel slippage, which meant it could climb the outer slopes of many craters. The original prototypes were larger, and the final unit actually become much smaller, which is highly unusual. No one has built a rover of this size capable of transporting a payload of this size at these speeds since the LRV was built. This is a challenge!

5.3 Field Work in a Space Suit

How did the astronauts perform field exploration in a spacesuit? Although the astronauts would have to perform field work that would involve using their hands, with a heavy emphasis on sampling, they would be encapsulated in suits with arms and gloves of restricted flexibility at the joints. What did that mean in terms of spacesuit, especially glove, design? An unforeseen component of the environment, lunar dust with the character of 'abrasive Velcro', aggravated that problem [10]. Fine dust accumulated in every joint and could not be brushed away. How were the geological field tools modified for use in spacesuits, and in a dusty, lower gravity environment? What did that mean in terms of method of documentation? These are the topics of this chapter.

5.4 Space Suit Design

Great efforts were made to provide a spacesuit [11–13] (Fig. 5.3) with manual dexterity: gloves were custom made for each astronaut's hands using plaster casts and then rubber molds, forming the innermost pressure retaining envelope, and incorporating silicone tips to provide 'touch' sensitivity as well as strength. However, the gauntlet-like gloves (Fig. 5.4) were still relatively stiff and cumbersome.

- The interior layer, in contact with the astronaut's hands, was 'a dipped neoprene/latex blend with impregnated nylon tricot knit' (glove paper downloaded), which had protrusions for the knuckles on the back of the hand. The cast of the hand had been made with the hand in a semi-clenched

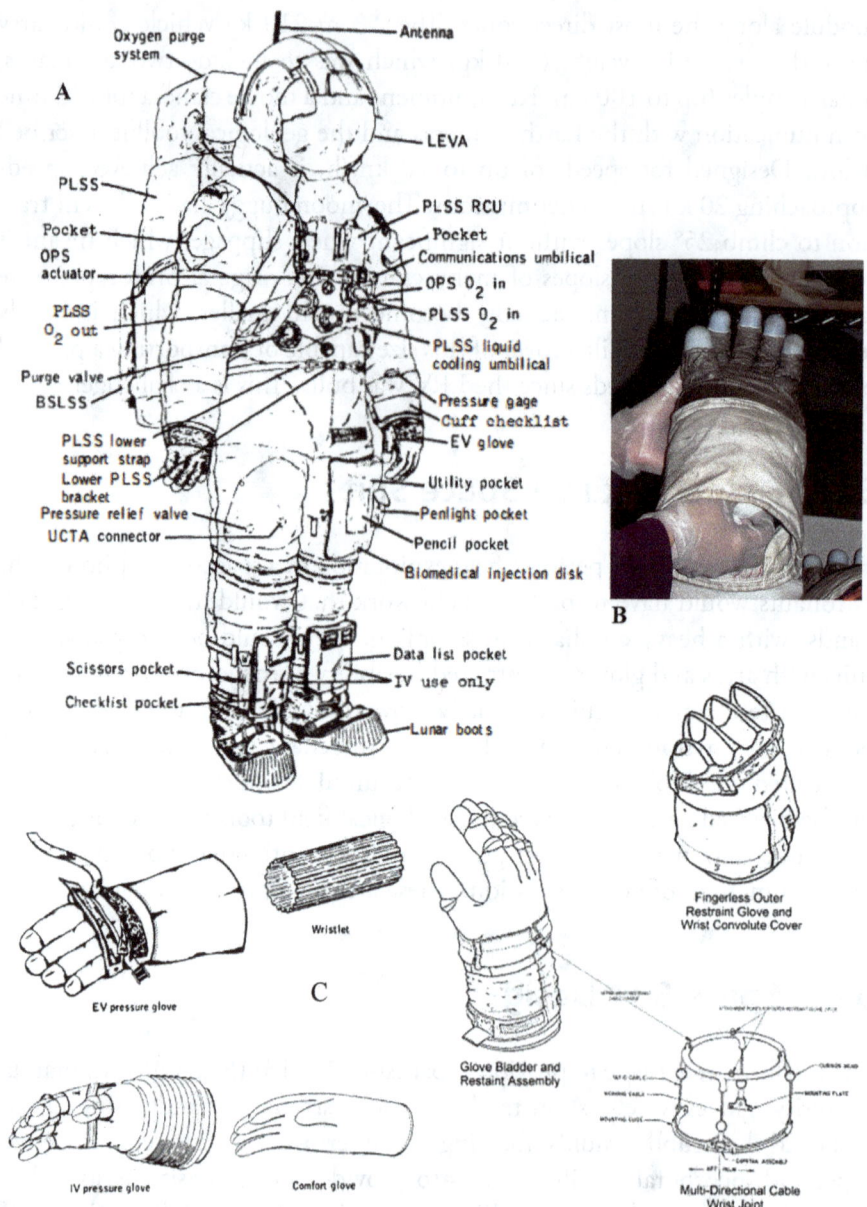

Fig. 5.3 (**A**) Apollo spacesuit with EVA features [11]. (**B**) Apollo 16 astronaut Duke's glove [11]. Note Thermal Micrometeorite Shield and reinforced fingertips [14]. (**C**) Glove components: Astronaut puts on Comfort glove, then IV pressure glove with bladder and inner restraint, then EV outer glove with convolute wristlet (to allow movement) and shield (seen in B) [12, 13]. (Courtesy of NASA)

position, which made further clenching somewhat easier, but opening far more difficult. The wrist joint allowed movement along two axes, including abduction, flexion/extension, and rotation.
- A neoprene/latex sheath, the next layer, provided added protection and anchored the palm restraint, a webbing enclosed wire across the palm through the thumb crotch which improved the astronaut's grasping capability and could be adjusted by the astronaut via a flap-covered black nylon strap with a Velcro closure on the back of the glove.
- The outer layer consisted of silicone-coated chromel-r, a woven stainless steel alloy fabric made with 20% chromium ad 80% nickel wires, to provide resistance against abrasion, wrapped around the thermal insulation. The insulation was composed of seven layers of aluminum-coated mylar separated by non-woven dacron spacers. A tacky palm surface on the outside of the glove was provided by a urethane coating.

5.5 Field Documents and Documentation

The indispensable geological field notebook, requiring use of a writing implement and both hands, was not practical for a variety of reasons. NASA's priorities were that the astronaut's suit extremities, gloves and boots, had to retain pressure in a vacuum, resist severe abrasion of the lunar regolith with which they were in constant contact, and provide protection against extremes of heat and cold.

Nevertheless, the constant close gripping of a writing implement would have put a great deal of strain on the astronauts' hands. In addition, at field sites the astronauts needed their hands for sample collection and photographic documentation. So, NASA came up with an alternative: articulate streaming of detailed yet comprehensible information, in other words, oral documentation (Table 5.9). In order to be clearly understood by everyone, descriptions would have to be very systematic, logical, highly descriptive, open-ended, and with minimal jargon. The astronaut narrative would be received on the ground in real time, transcribed, and analyzed by a geological 'backroom'. This process is described in the video 'when the astronaut speaks' [3]. This was a highly disciplined process: thus training would be required and would play the crucial role in documentation.

In fact, systematic oral description of sites and samples, recorded live and critiqued afterward) was a critical part of astronaut field training, occurring for every field trip. Such documentation, accompanied with panoramic/portrait photos to confirm position and context of collected samples numbered

Table 5.9 Systematic oral documentation of lunar surface geology

Type 1: Object Named	Type of Object (e.g., boulder) and general composition (e.g., anorthositic) provisional description		
Type 4	Contacts with other objects (e.g., conformable, intrusive, parallel, slumped, volcanic		
Type 2. Properties	Class A: optical properties	Examples include color, luster, albedo, texture	
	Class B: Mechanical properties	Examples include hardness, strength, coherence, resistance, geomorphic exposure	
	Class C: Geometrical properties	(1) Size-shape	e.g., thickness, angularity, dimension, extent tabularity
		(2) Orientation	e.g., altitude, direction, trend, position, location
	Class D: Structural properties	(1) Internal structures	e.g., fabric, cleavage, layering, texture
		(2) External structures	e.g., folds, faults, fractures, joints
Type 3	Final summation (object name confirmed specific name and composition)		

for later use formed the basis of documentation for every observation made while the astronauts were on the moon, upon landing, when traversing a field area, and at field sites during the Apollo missions [15]. In the verbal descriptions, each part of the scene would be described in a systematic order, from larger system to components, and then from components to smaller components, repeated at each level, as the scene was naturally being approached from a distance [16] (Table 5.8). Properties (visual, mechanical, structural, and geometrical) would be considered in the same order at each level of the description Furthermore, the level of detail would be indicated at the beginning, with relatively frequent pauses during the exposition, to facilitate the capture of information from audio streaming. Usage of terms would be agreed upon in advance, and comparison to familiar or already described objects encouraged. Such descriptions would be accompanied by visual images to portray complex relationships quickly. Verbal, as well as photographic, cues would indicate the rationale for selecting the sample and its relationship to the character of and objectives for exploring the station or sampling site. The description would be ended with an indication of the direction and rationale for the next sampling activity. To be most useful, descriptions include a context for the details of each station, as illustrated in Figs. 5.4, 5.5, and 5.6. In the Apollo 15 excerpt (Fig. 5.4), Scott and Irwin are driving to Mt. Hadley. The description provides a panoramic view as a context for the description of

122:36:30 Scott: Oh, look back there, Jim! Look at that. Oh, look at that! Isn't that something? We're up on a slope, Joe, and we're looking back down into the valley and...
122:36:39 Irwin: That's beautiful.
122:36:40 Scott: ...That is spectacular! (Pause) (I'll) get the (TV) antenna pointed here.*[Photos AS15-85- [11448](#) to [11454](#) are part of a pan Jim starts at [123:17:16](#)]*
122:36:51 Allen: Okay, Jim. And could you give us a frame count...
122:36:53 Irwin: I'm taking a pan.
122:36:54 Allen: ...when you finish your pan; and, Dave, we'd like one from you.
122:36:59 Scott: Yes, sir.
122:38:33 Scott: Okay. Going to [FM/TV](#).
122:38:41 Scott: Man, you all ought to have a great view this time. Okay, Jim; let's go sample this rock.
122:38:46 Allen: Can hardly wait.
122:38:47 Irwin: Let me take a pan here, Dave. (station 2 pan, AS15-85-11422 to 11438
122:38:48 Scott: Okay; get your pan. (Responding to Joe) This is unreal. *[TV on.]*
Video Clip 2 min 46 sec ([0.7 Mb RealVideo](#) or [25 Mb MPG](#))
122:38:53 Scott: The most beautiful thing I've ever seen. (Long Pause) Man, we're walking uphill, too! Is that ever uphill! There is one boulder! Very angular, very rough surface texture. Looks like it's partially...Well, it's got glass on one side of it with lots of bubbles; and they're about a centimeter across. And one corner of it has got all this glass covering on it; seems like there's a linear fracture through one side. It almost looks like that might be a contact; it is, within the rock. It looks like we have maybe a breccia on top of a crystalline rock. It's sort of covered with glass; I can't really tell. But I can see a definite linear feature through one side of it which is about a fifth, and the glass covers both sides of what I guess I'm calling a contact.
122:40:28 Scott: And there's also - parallel to that contact - one surface which is quite flat, (although) only for about 8 inches or so. Looks like it's been chipped off. The boulder itself is on the order of about a meter across and maybe....Gee, it looks like a half meter thick or so. It's got a fillet up one side, and the other side is in a shadow. I can't really tell whether - It doesn't look like it's filled. It's got a fillet on the downslope side, and the upslope side is open and free. As a matter of fact, it looks like
122:41:04 Irwin: It looks fairly recent, doesn't it, Dave?
122:41:07 Scott: Yeah, it sure does! It sure does, and I can see underneath the upslope side; whereas, on the downslope side, it (the soil)'s piled up. Boy, that is really something. Hey, let's get some good pictures of that before we disturb it too much.
122:41:25 Allen: Roger, Dave and Jim...
122:41:26 Irwin: Do you want a sample ...
122:41:28 Allen: ...See you crystal clear, and we've got a beautiful tally-ho on you and boulder on the TV.
Video Clip 2 min 25 sec ([0.6 Mb RealVideo](#) or [22 Mb MPG](#))
122:41:31 Allen: And it probably is fresh; probably...
122:41:34 Scott: Okay.
122:41:35 Allen: ...not older than three and a half billion years.
122:41:41 Scott: Can you imagine that, Joe? Here sits this rock, and it's been here since before creatures roamed the sea in our little Earth

Fig. 5.4 Apollo 15 Surface Journal Excerpt [17]. (Courtesy of NASA)

166:47:14 Young: Man, does this thing have steep walls.
166:47:16 Duke: They said 60 degrees.
166:47:18 Young: Now, I tell you, I can't see to the bottom of it, and I'm just as close to the edge as I'm going to get.
Video Clip (3 min 13 sec 0.8 Mb RealVideo or 28 Mb MPEG)
166:47:44 Young: Okay. Our plan here is to range along the edge of the crater for about 80 meters, if you can do that. And, Charlie, if you'll start out with your pan 1, then the 500 millimeter. We'd also like you to shoot some more pictures of Smoky with the 500, and then take your far field stereo. And then if you range on out as far as you can go, taking the 500 millimeter with you, take - and John with you - take a stereo of the inside of the crater with the 500 millimeter from as far away from the Rover as you can get. And then stick the 500 millimeter in John's SCB, and then do your other far-field polarimetry, and then, from then on, all we've got is sampling.
166:49:06 England: And we've got a picture. (Pause)
166:49:13 Duke: Okay. Those rocks you're looking at now, Tony, are white and they look breccious to me. The big black one (meaning House Rock) is off behind the TV. And you're going towards the rim on the crater right now
166:49:34 Young: The unfortunate thing about it, Houston, is that rascally rim...It goes down...It slopes into it about, say, 10 or 15 degrees, which is the kind of slope I'm standing on right now; and then, all of a sudden, in order to see to the bottom, I've got to walk another 100 yards down a 25- to 30-degree slope, and I don't think I'd better. Maybe we can drive around to the other side and see down into it.
166:50:13 England: Man, is that a hole in the ground!
166:50:14 Duke: Okay, Tony. The inside...(Stops to listen) It really is. I see no bedrock, though. All I see is boulders around the crater. There's nothing that reminds me of bedding, just loose boulders. Though it might very well be that it's so shocked
166:50:37 Young: Now, the layering, the boulder layers are horizontally oriented; and, of course, they are all covered with talus. Over on the north wall, in particular, about 1/3 of the way from the top is a line of boulders that you probably ought to be able to see on the TV. But they're all oriented right in that line, which would lead one to think that it has bedding there. Don't you see that line right over there, Charlie?
Video Clip (3 min 10 sec 0.8 Mb RealVideo or 28 Mb MPEG)
166:51:11 Duke: I don't...I'm worrying about trying to get this crazy (500-mm) camera going here.
166:51:17 Young: Okay. And the...(Pause)
166:51:24 England: Okay...
166:51:24 Young: The material is just completely different...The wall...(Stops to listen)
166:51:25 England: ...That line of boulders on the north wall, what color were they?
166:51:30 Young: In this light, they appear to be dark boulders.
166:51:36 England: Okay. Incidentally, the white rocks you see there. Do they look like the Cone-Crater-type white rocks? (Pause)
166:51:48 Duke: No, not to me.
166:51:51 England: Okay.
166:51:55 Young: Better let me get a piece of one, Charlie. I don't think...This is definitely a breccia right here, a big foot-and-a-half breccia. It's a white matrix with dark clasts, and it looks to be a three-rock breccia. Some of the dark clasts have even darker clasts than those. (Pause)
166:52:24 Duke: Okay, Tony. (Before leaving the Rover) I picked up (probably means "loaded into the camera") magazine Mike; it's on the 500.
166:52:29 England: Okay.
166:52:30 Young: Okay. Houston, I just picked up a grab sample of a breccia. It's very friable. It looks shocked; it has black clasts in it, black clasts (which are) a couple of millimeters across. (Pause) It's so worn down that you know what it really looks like? It looks like a...If I can use the analogy, (because) I'm not sure what the heck it is. It looks like a tuff. It just looks like a rock with a...You see, the clasts are sticking out of it, is what I'm saying.
166:53:24 Duke: (At 166:47:44, did) you say you want Smoky or Stone Mountain?
166:53:29 England: Okay. We'd like some more pictures of Smoky.
166:53:38 Duke: Okay.
166:53:41 England: And, John,...
166:53:43 Duke: Get up here where I can see it (meaning Smoky Mountain).
166:53:43 England: ...in your mineral description there, could you see a crystal shape?
166:53:44 Young: Okay. While Charlie's doing... (Stops to listen)
166:53:46 Young: "Can I see a crystal shape?" I saw one clast. One...No. (Pause) Well, the clasts in there are very angular. Maybe that's a zap crater; that's probably what that is.
166:54:07 Young: I don't see in the...The white matrix doesn't have any crystalline structure that I can recognize.

Fig. 5.5 Apollo 16 Surface Journal Excerpt [18]. (Courtesy of NASA)

142:52:53 Schmitt: Okay, Bob. The blue-gray rocks are breccias. They're multilithic, gray-matrix (pause), matrix (dominated) breccias, I guess. There are fragments in them, but it doesn't look like more than about 10 or 15 percent fragments.

142:53:10 Schmitt: Some of the light-colored fragments seem to have very fine-grained dark halos around them. The zap pits (in the dark matrix) do not have white halos, so I suspect they are not crystalline (rocks). They might be the vitric or glassy breccias. At least, the one big rock we have here.

142:53:45 Schmitt: There's a very rough foliation in them (the blue-gray breccias) - and I'm not sure - it's shown by the elongate knobs on the surface. It looks like a fracture foliation of some kind.

142:54:00 Cernan: Jack, that rock has almost got to have come down (from the outcrops high on the mountain), don't you think?

142:54:04 Schmitt: Oh, no question about it. I'll bet you it's the same as the blue-gray rocks we see up higher. Here's some more blue-gray ones over here.

142:54:11 Cernan: Let's start taking...Oh, yeah. Look at the size of some of these light fragments in here.

142:54:18 Schmitt: Yeah, but it still...It looks like they're dominantly matrix breccias. There are light-colored fragments, and they (the fragments) may be crystalline.

[Fendell finishes his counter-clockwise pan at the Nansen/South Massif contact and reverses direction.]

142:54:34 Cernan: Okay...

142:54:35 Schmitt: They are. They're very light colored; they look like the shattered anorthosites. They have white halos (around zap pits)...I think that's what those fragments are.

Fig. 3 Example of application of systematic description technique in script from Apollo 17 (Apollo 17 Lunar Surface Journal)

Fig. 5.6 Apollo 17 Surface Journal Excerpt [19]. (Courtesy of NASA)

the boulders at the site as a prelude to sampling. In the Apollo 16 excerpt (Fig. 5.5), Young and Duke are at North Ray Crater. First, they describe the crater itself, then the rock wall exposures, then details of individual boulders. In the Apollo 17 excerpt (Fig. 5.6), Schmitt and Cernan are at Nansen Crater (station 2). Schmitt is describing a rock, starting with an overall description of color and texture, and then describing individual components.

5.6 Documentation Devices: Maps and Cuff Checklists

Starting in 1969, the USGS would produce photogeological maps with traverses indicated on aerial photos, as described above. Marked on the map were major field sites (stops), distance between stopes, timelines, and compass headings. Each geological unit indicated on the map had a description which was based on what could be inferred using standard photogeology techniques in which the astronauts had been trained. The USGS also produced space suit

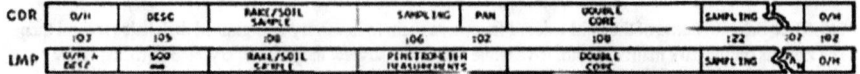
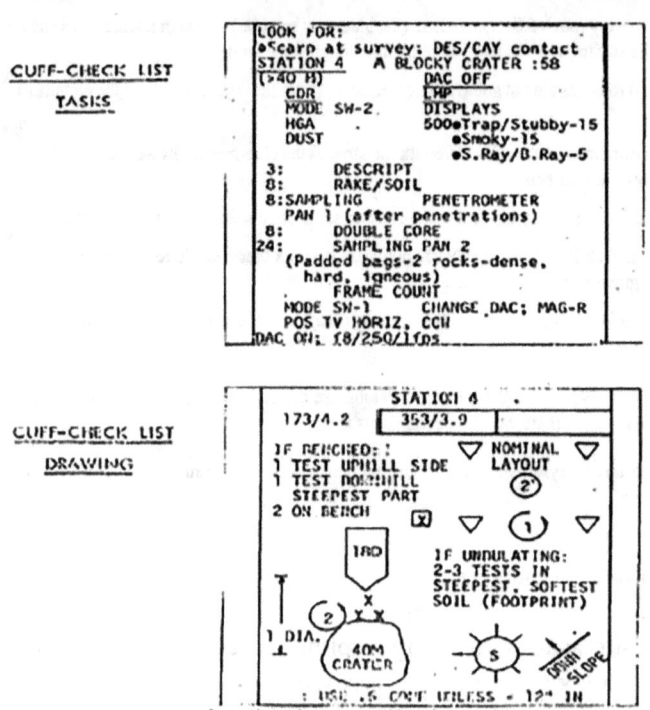

Fig. 5.7 Schedule and cuff check list for Apollo 16 Station 4 Stone Mountain. Note nature of schedule and mnemonic nature of checklist as described in text [18]. (Courtesy of NASA)

cuff checklists (Fig. 5.7). Cuff checklists were little spiral bound notebooks held in place with long Velcro straps placed over the cuff of the spacesuit (Fig. 5.8). Pages were devoted to critical activities during experiment package deployment and geological traverses, as well as resolving life-threatening problems with the EMU (extravehicular mobility unit: the spacesuit). The pages included concise, mnemonic instructions and drawings.

The field exercises were conducted using aerial photomaps produced at the same scale as the photomaps to be used on the Moon (Figs. 5.9, 5.10, 5.11, and 5.12). The maps displayed geological units delineated by textures and shades that suggested these units, just as they would be on the lunar photomaps. Traverses and station locations, as well as notable topographic features,

Fig. 5.8 Apollo 17 astronaut Cernan with cuff check list, closeup on the right. (Courtesy of NASA)

Fig. 5.9 Apollo 14 Astronaut Mitchell using map. (Courtesy of NASA)

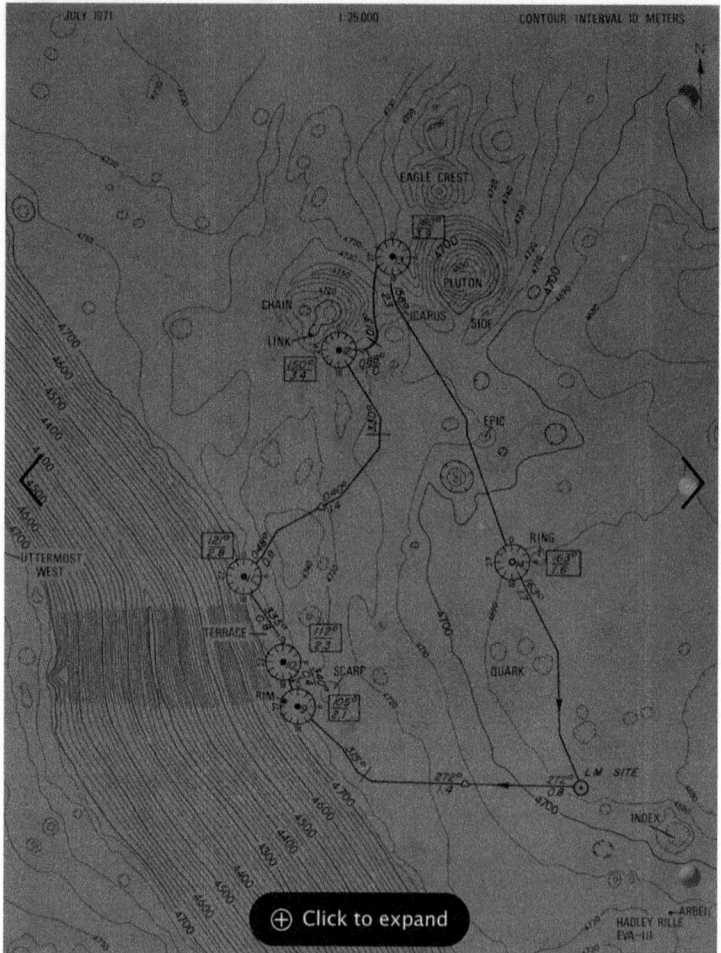

Fig. 5.10 Apollo 15 Lunar Navigation map used on lunar surface. (Courtesy of NASA)

were marked as well. As noted in the last chapter, the training for each individual Apollo mission became more and more like the real mission during the last few months before each flight, and the final field exercise was conducted in a terrain similar to that of the actual landing site. During the exercise, the astronauts communicated directly with the personnel manning the consoles at Mission Control in Houston.

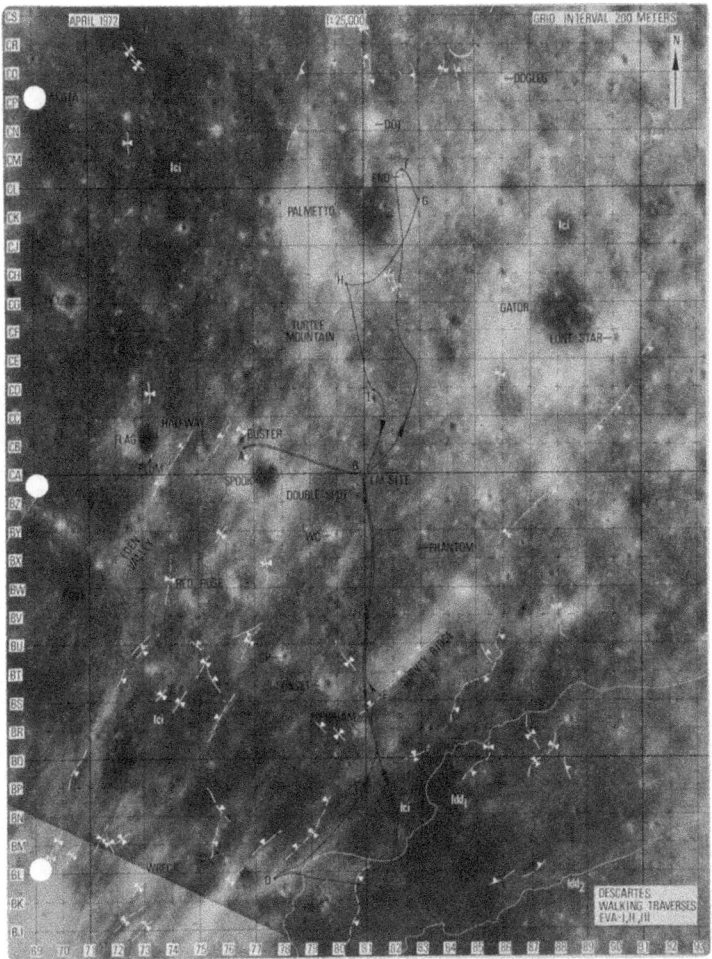

Fig. 5.11 Apollo 16 Lunar Navigation map used on lunar surface. (Courtesy of NASA)

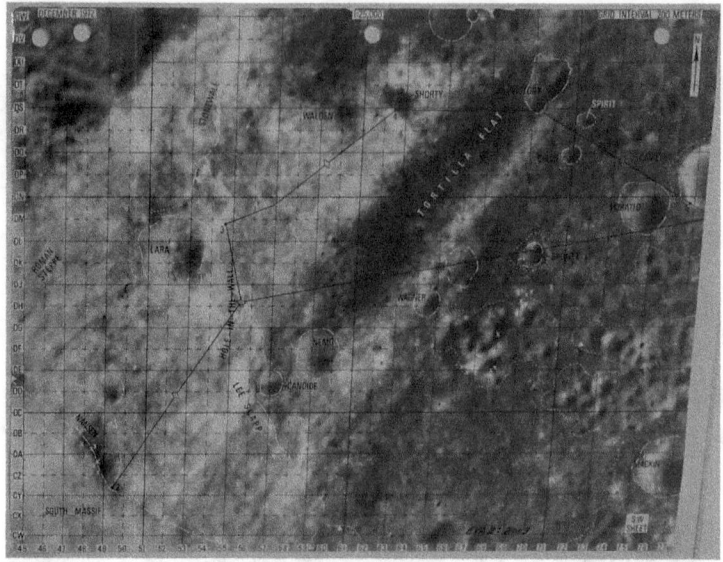

Fig. 5.12 Apollo 17 Lunar Navigation map used on lunar surface. (Courtesy of NASA)

5.7 Field Work Practices

Field work involved planning as well as traverse activities. They learned to study a site through the use of photomaps, to anticipate keys to the site's history and potential sampling sites. Early on, they learned to find the sites of interest and follow traverses that had been developed by the geology team. As their geology skills improved, and they learned how to use the geological toolkits described below, they learned to design traverses from sites they themselves identified which would provide samples representing major events, provided a context and understanding of processes that shaped the training site. They learned to determine the most effective type of soil sampling to use, including bagging, trenching, or raking soils, or driving core tubes. They learned to provide photo-documentation of sites and samples at the appropriate scales, including overlapping panoramic shots of the surroundings, photos of sampling sites, and close-ups of samples themselves. In addition, they set up experiment packages and took geophysical measurements. During all of these activities, they recorded measurement, sample, or photo locations and numbers. Meanwhile, the geology team followed the traverse, and, through verbal descriptions, a staff-mounted camera, and video imagery, produced geological maps.

References

1. Fishman, C. One giant leap: the impossible mission the flew us to the Moon. Simon and Schuster, 480 p., 2019.
2. Belbin, Meredith. Management teams: why they succeed or fail. 3rd ed., Taylor and Francis, 208 p., 2010.
3. USGS Astrogeology Branch, Astroshare Videos, Apollo Astronaut Training Films (originally 16 mm and translated into mp4 format), edited by Marc Hunter, 1968-72.
4. Kiesman, A. Apollo astronauts almost got to fly lunar jetpacks. Astronomy Magazine. May 31, 2019. https://www.astronomy.com/space-exploration/apollo-astronauts-almost-got-to-fly-lunar-jetpacks/. visited 10/1/24
5. NSSDC, The Apollo Lunar Roving Vehicle. https://nssdc.gsfc.nasa.gov/planetary/lunar/apollo_lrv.html. visited 10/1/24
6. Baker, D., Lunar Orving Vehicle: Design Report, Spaceflight, 13, 234–240, 1971.
7. Burkhalter, B. and M. Sharpe, Lunar Roving Vehicle: Historical orign,s development, and deployments. Journal of the British Interplanetary Society, 48, 199-21, 1995.
8. Wright, M., B. Jaques, S. Morea. A brief history of the lunar roving vehicle, as part of the history of the NASA Marshall Space Flight Center. NASA, 27 p., 2002. https://www.nasa.gov/wp-content/uploads/static/history/alsj/msfc-lrv.pdf
9. McMillan, D., Lunar Rover Vehicle Foldup Animation. NASA Video. 2013 https://www.youtube.com/watch?v=7OL3OmM-CYQ. visited 10/1/24
10. Clark, P.E., S. A. Curtis, F. Minetto, J. Marshall, J. Nuth. SPARCLE: Electrostatic Dust Control Tool Proof of Concept, SPESIF-10, Ed. G. Patterson, AIP Conference Proceedings, 1208, 541–556
11. Johnson Space Center Crew Systems Division, Apollo Operations Handbook Extravehicular Mobility Unit, NASAMSC-01372-1, 1968. https://www.nasa.gov/wp-content/uploads/static/history/alsj/alsj-EMU1.pdf. visited 10/1/24
12. Ayrey, W., Chapter 1: The Path leading to Space, in ILC Spacesuits and related products 0000-712731, Rev. A, 2007. https://www.nasa.gov/wp-content/uploads/static/history/alsj/ILC-SpaceSuits-RevA.pdf
13. Wright, H., Enhancement of Space Suit Glove Performance. SAE Transactions, 94, 5, 210-225, 1985.
14. Jones, E., and K. Glover, Apollo Lunar Surface Journal, Apollo 16, Charlie Duke's Glove. https://www.nasa.gov/history/alsj/a16/A16CMDLeftEVGlove.html. visited 10/1/24.
15. Clark, P.E., Revolution in Field Science: Apollo approach to inaccessible surface exploration. Earth, Moon, and Planets, 106, 2–4, 135–157. http://www.springerlink.com/content/uk63tln642601617/

16. M'Gonigle, J., D. Schleicher, and I. Lucchitta, A proposed scheme fo rlunar geologic description, UGSG Interagency report: Astrogeology 18, 1969.
17. Jones, E., and K. Glover, Apollo Lunar Surface Journal, Apollo 15. https://www.nasa.gov/history/alsj/a15/a15.html. visited 10/1/24.
18. Jones, E., and K. Glover, Apollo Lunar Surface Journal, Apollo 16. https://www.nasa.gov/history/alsj/a16/a16.html. visited 10/1/24.
19. Jones, E., and K. Glover, Apollo Lunar Surface Journal, Apollo 17. https://www.nasa.gov/history/alsj/a17/a17.html. visited 10/1/24.

6

Field Work Tools

6.1 Sampling: The Primary Field Work

Sampling accompanied by documentation was the primary geological activity. Between major stations, traverse samples would be collected on the scale of tens to hundreds of meters, sometimes without even getting out of the rover. Sampling was a systematic yet flexible activity. Stations along the route were selected in advance on the basis of the apparent availability of exposures of underlying strata, typically boulders [1]. Even though sampling activities at a given station were typically described in terms of allotted time and site geology in the planning documents, a pattern in sampling activities is discernable and a technique for selecting sampling sites within a station is implicit in the Apollo Lunar Surface Journals [2, 3].

Samples were collected for return to Earth using the various implements described herein. Particulate samples were always placed in numbered bags and ultimately bags were placed within boxes. Rock samples were sometimes placed in bags before being placed in containers called rock boxes. Planners and trainers briefed the astronauts on the geological traverses and stations, describing the geology and their working hypotheses, identifying potential sampling sites that would allow the hypotheses to be tested [4]. Then the astronauts participated in simulations of the traverses. Debriefing sessions with all present were scheduled after the simulations as well.

The two Apollo astronauts worked as a team in the field, and interacted with the geology 'backroom' in near real time as well [3]. These skilled were developed and practiced during field site training. They might get feedback as to where to go next or further questions on a sample they had just acquired.

By definition, astronauts were trained to select a limited number of representative samples (rocks or regolith), and thus perform a built in 'down select' based on visual cues linking the surrounding landscape to only certain among all possible samples. Astronauts had also been trained to minimize the time required for sample collection during their training, which involved time motion studies [4]. One astronaut might take the pictures for documentation while the other one gathered the sample. One astronaut might be taking a measurement while the other deployed equipment. Astronaut performances improved with each EVA and with each mission as the field methodology matured. At a typical station, 3–4 prime sampling sites were reachable on foot from the rover, separated by tens of meters [1]. Panoramic photographic documentation was provided at the very beginning of time spent at a station. Verbal and photographic descriptions were provided on the way to sampling sites, and then the sites were systematically sampled [3, 5, 6].

Site activities typically involved systematic sampling of soils with small rock fragments around and under a boulder as well as using the hammer to obtain chips of the boulder or collecting larger rocks below that apparently originated from it. A variety of tools were used, to sample surface and subsurface soil and rock components with a variety of dimensions. Shovels and scoops typically sampled soils with existing small rock fragments, rakes captured fragments on the order of 1 mm or greater, and larger fragments were gathered by tongs, by hand, or were chipped off of a boulder with a hammer. The shovel was also used as a trenching tool to observe the properties, including the distribution of rock fragments, of the regolith near the surface. Two types of core tubes were used to collect profiles of the regolith. Tools are described below. Collecting the deep drill cores took considerably more effort. As pressure is applied to the regolith from the top, the regolith compacts, becoming harder and considerably more difficult to penetrate. The drill bit was modified after the Apollo 15 mission, in an effort to make penetration easier. All sampling devices and containers are described below.

The amount of time devoted to each sample was typically on the order of 2–3 min (except in cases of the deep drill cores). Table 6.1 provides further description of these activities [1, 5, 6]. During the course of each of the day's traverse, astronauts typically gathered 30–35 kg of samples from 4 to 5 major stations, averaging between 7 and 8 kg per station. Astronauts had been trained to minimize the time required for sample collection during their training, which involved time motion studies [4, 7–9]. At a typical station, 3–4 prime sampling sites were reachable on foot from the rover, separated by tens of meters. Sampling at a site typically took 15–20 min and was repeated for each sampling site [1, 5, 6]. Systematic sampling typically involved sampling

Table 6.1 Time and resource commitments for lunar surface activities

Activity	Pre-activity	During activity	Post-activity
Site characterization 5 min/event, 2 crew	Panoramic view (audio and photo) to locate sample sites, outcrops		
Surface sampling, soil, 2–3 min/sample, 1 crew	Orient, document (audio and photo)	Collect using shovel or scoop w/or w/out long handle	Document (audio and photo), encapsulation in bags
Surface sampling, fragments, 2–3 min/sample, 1 crew	Orient, document (audio and photo)	Collect using rake	Document (audio and photo), encapsulation in bags
Surface sampling, rock, 2–3 min/sample, 1 crew	Orient, document (audio and photo)	Collect using glove, tongs, rock hammer and chisel	Document (audio and photo), encapsulation in bags
Subsurface sampling (20 cm) shallow core, 4 min/sample, 1 crew	Orient, document (audio and photo)	Collect using core tube or shovel/trench tool	Document (audio and photo), encapsulation in bags
Subsurface sampling (2 m) regolith drill, 25 min/2 m, 1 crew plus last 5 min, 2 crew	Orient, document (audio and photo)	Power drill	Document (audio and photo), encapsulation in bags
Crustal structure, gravimeter, 5 min/reading, 1 crew	Orient, document (audio and photo), set up gravimeter	Take gravimeter measurement for station	Stow for next station
Crustal structure, active seismic, 20 min plus 3 min/explosive device, 1 crew	Orient, document (audio and photo) site for central station and geophones	Deploy central station and geophones near lander on foot	Deploy explosive devices at selected locations with underlying structure of interest
Crustal structure, magnetometer, 1 crew, 30 s	Orient, document (audio and photo), set up magnetometer	Take magnetometer measurements for station	Stow for next station
ALSEP package, 2 crew, 3 h	Orient, document (audio and photo) site for central station	Set up central station and individual experiments, connect to central station	Test and turn on (removing dust covers) when departed

soils and rock fragments around and under a boulder as well as obtaining chips of the boulder or rocks on the ground that apparently originated from it.

The Apollo astronauts used the unpressurized rover to travel distances of up to 25–30 km during the course of a traverse during an EVA. They spent about half of their time at the stations, and the rest driving. This pattern was developed during their training with a simulated rover [1, 10].

Other important non-sampling activities are described in Table 6.1. The active seismic experiment was deployed during the mission, requiring 20 min for one astronaut to deploy the central station and geophones near the lander on the first day. On following days, a series explosive charges, up to 8 on Apollo 17, were deployed on traverses over potentially interesting underlying structures, taking up to a few minutes apiece to deploy Gravimeter and/or magnetometer measurements were taken at many of the stations for missions on which they were flown, requiring about 5 min each time the measurements were taken.

6.2 The Toolkit

Fortunately, geological surface sample collection tools shown in Table 6.2 [11] didn't require fine scale manual dexterity, and in fact were optimized for use in the gloved hand of the astronaut. Designs were improved through the use of time and motion studies as astronauts used tools in the field during their training [2, 4–6]. Tools, carried in a tool carrier, are cataloged and illustrated in Figures below.

From Apollo 12 on, the astronauts had tool carriers, a smaller hand- carried one for Apollo 12 and 14, and a larger rover mounted one for the J missions (Fig. 6.1).

Surface sampling handling tools included:

- Scoops and extension handles. Scoops were used for the collection of representative regolith samples. The size of scoop, along with the length and manner in which the handle could be extended was also a matter for special consideration. A scoop with a handle extendable over a range of angles and lengths, was specifically designed for quickly obtaining a sample from the rover without dismounting. (Fig. 6.2)
- Hammers were used for obtaining representative samples of larger rocks or boulders (Fig. 6.3)
- Tongs were used for picking up and delivering rocks to sample containers {(Fig. 6.3)

Table 6.2 Apollo Geological Toolkit

Type	Tool	~Mass (kg)	Support objectives requiring
Sample Collection	Rock Hammer and Chisel	1.5	Rock collection
	Rake	1.5	Rock collection
	2 Tongs	1	Rock collection
	Shovel	1.5	Soil collection
	Rover Sampler Scoop	0.25	Soil collection
	Large Scoop	0.5	Soil collection
	2 adjustable extension handles	1.5	Soil collection
	Tool Carrier	10	Sample collection
	Regolith deep drill cores (1–2 m) w/drill stems	13	Subsurface sampling
	Shallow core drive tubes (20–30 cm)	5	Subsurface sampling
	Sample Bags	10	Soil collection
	Rock Boxes	15	Rock collection
	Scale	0.25	Sample collection
Site Characterization	Photographic Equipment	25	Site and sample characterization
	Gnomon	0.25	Size, scale, and lighting conditions
	Magnetometer	10	Surface magnetic field variations
	Gravimeter	15	Variations in mass and underlying structure
	Active Seismic Experiment	10	Underlying structure to depth of tens of meters

- The shovel was an adjustable angle trenching tool on Apollo 14, replaced by the adjustable angle scoops on later missions (Fig. 6.3)
- The rake was used to 'sieve' a sample of regolith for 'pebbles' (>1 cm size) (Fig. 6.2)

The astronauts used a variety of sample containers (Figs. 6.4 and 6.5).

- Dispensers made the smaller bags (all 8 cm deep cup shaped Teflon with sealable aluminum rims on the J missions) readily available to the astronauts. These bags were used for unconsolidated regolith samples collected via scoops, rakes, or shovels.
- Padded (15 × 14 × 5 cm) sample bags were available for rocks.
- Sealed 'Special Environmental Sample Bags' were also available for particularly valuable samples.

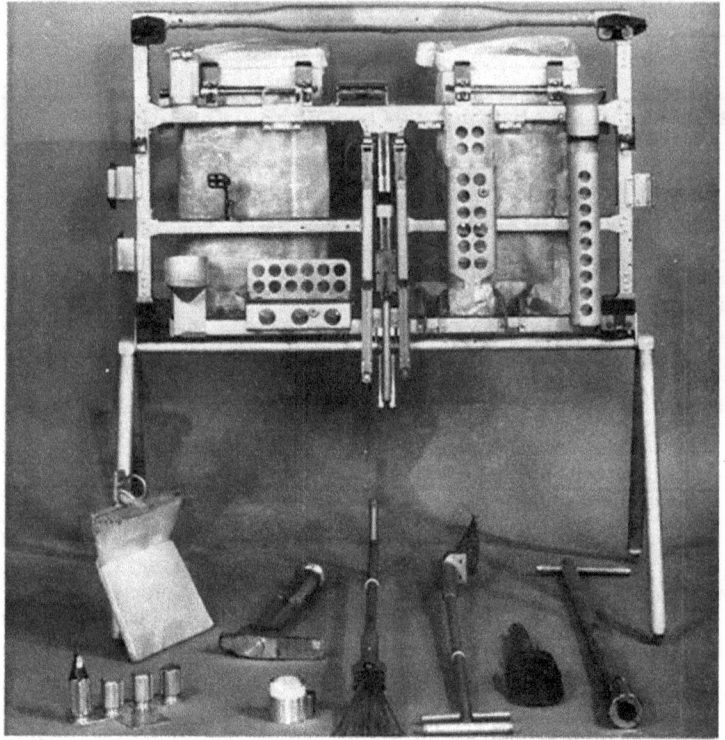

Fig. 6.1 Large tool carrier used on Apollo J missions with some tools displayed below left to right: core tube caps, documented sample bags, hammer, drive tube caps, two pairs of tongs, adjustable handle scoop and extension handle [11]. (Courtesy of NASA JSC)

- Larger sample collection bags, attached to the astronaut's pack or the rover, were used to hold collected and labeled regolith or rock samples. Sample collection bags could also hold core tube samples, as described below.
- Finally, all samples were packed in hard-sided 'rock boxes' for delivery to Earth onboard the LM.

Other items useful for characterizing samples as they were collected.

- A gnomon (Fig. 6.5) with a color reference chart was used to indicate time of day and to identify sample colors under lunar conditions.
- A spring scale could be used to weigh the sample (Fig. 6.5).

A surveying staff to indicate distances was eliminated because it was awkward to carry. The lunar staff went through various design changes, but essentially

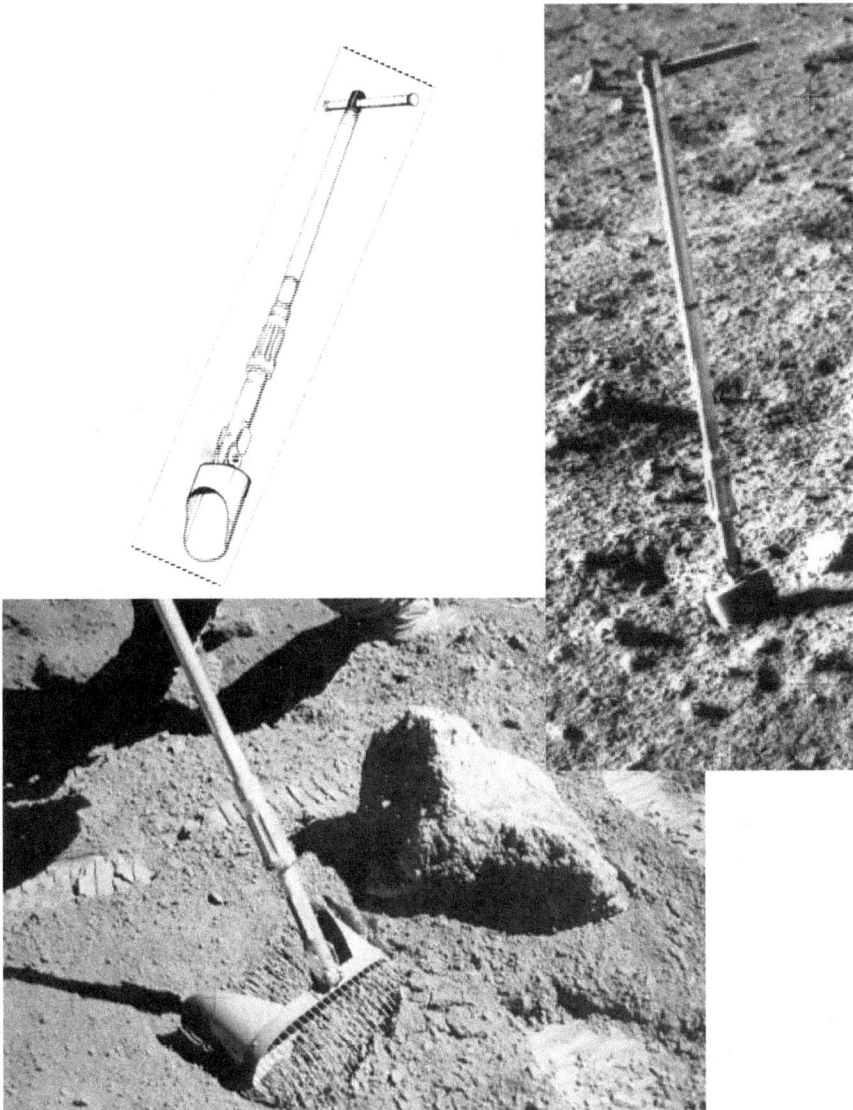

Fig. 6.2 Sample Collection Tools [11] clockwise from top: Small adjustable-angle scoop attached to long extension handle drawing (left) and being used during Apollo 16 (right). Bottom: Rake sample collecting rock fragments during Apollo 16. (Courtesy of NASA JSC)

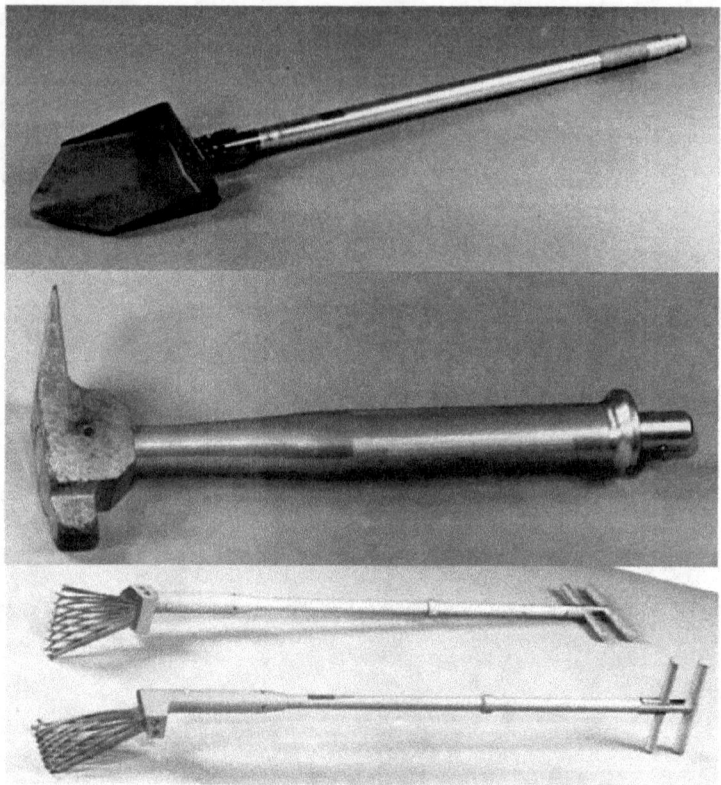

Fig. 6.3 Sample collection tools [11] from top to bottom: Trenching tool with adjustable angle blade, Apollo J mission heavier hammer, 32-in. tongs [11]. (Courtesy of NASA JSC)

would have included a mobile television camera, a laser range finding reflector system, a spectrographic system, a sun compass, the LGEC, and various other experiments for performing lunar geology on site (Fig. 6.6) (quoting from Karl Dodenhoff) [12]:

> The LGEC finally met it's demise due to two major problems. One was that "*the company*" (which I have failed to find a name for anywhere in the available literature) that NASA hired to develop it from prototype to a flight model was too small to handle the complexities of the project and subsequently went bankrupt. Also, there seems to have been some political intrigue behind this, as stated in the USGS paper: "The lunar stereo camera [that the Branch was developing and field testing in astronaut simulations] was cancelled because development was under Leonard Reiffel [in George Mueller's Office, Office of Manned Space Flight, NASA HQ], who never liked the camera, Gene Shoemaker, or the USGS

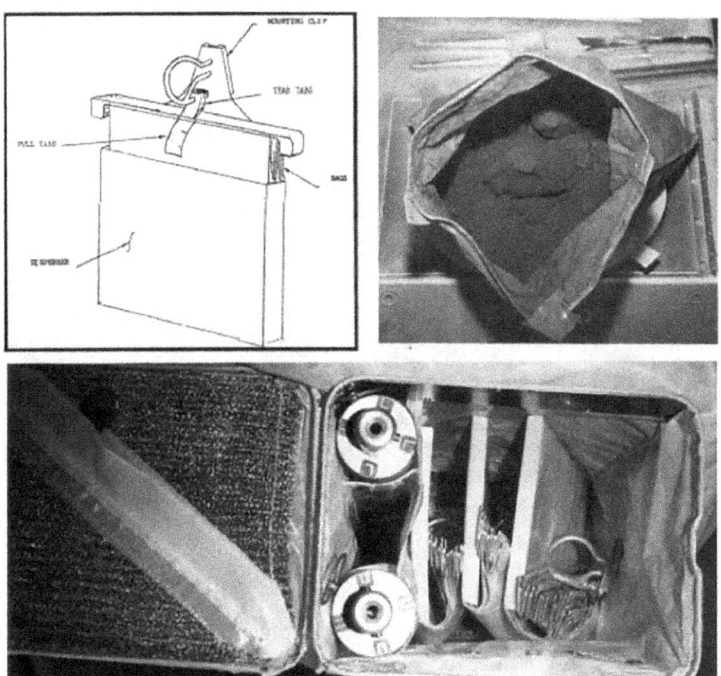

Fig. 6.4 Sample documentation equipment [11]. Top left 20-bag dispenser for flat, rectangular labeled bags on Apollo J missions. Right one of the bags opened in the lab to show Apollo soil sample 74,220. The aluminum rims hods bag open. Bottom: three 20 bag dispensers packed inside sample collection bag prior to flight. (Courtesy of NASA JSC)

for that matter." So, instead of using the LGEC, the astronauts were trained how to use their Hasselblads to take stereographic photo sets of geologic samples in situ.

Another proposed package that was tested during simulations but never flew on the Apollo missions due to mass, power, and time constraints was the desktop size automated thin section and slicing machine designed by Microtek (Fig. 6.7). The concept, an idea so compelling to lunar geologists that it wouldn't die, has been revisited and proposed in various forms several times since then, in 2007–2009 as 'suitcase science' as part of Lunar Sortie Science Opportunities program during the 'Project Constellation' Vision for Space Exploration promoted by former President George W. Bush.

The subsurface sampling devices ranged from manual shallow drive tubes (20–30 cm), with upper and lower tube sections (the lower section could be used separately or combined with the upper) (Fig. 6.7) to the powered

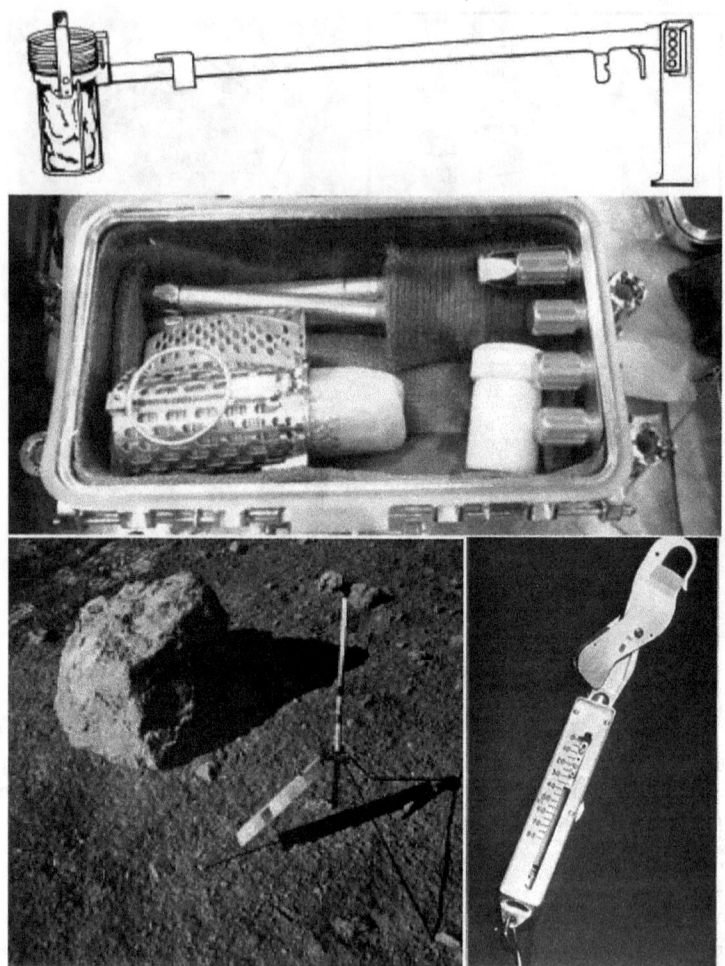

Fig. 6.5 Sampling equipment [11]. Top: Cup-shaped labeled sample bags used in LRV soild smaple on Apollo 17. Middle: Apollo 14 sample return container prior to flight with round documented sample bags, 2 cm diameter core tubes, core tube caps, and magnetic shield sample container (white cylinder). Bottom left Gnomon at Apollo 17 site, right sample scale. (Courtesy of NASA JSC)

rotary-percussive regolith drill (up to 3 m using several core stem sections) [3] (Fig. 6.8). These subsurface samplers proved to be more difficult to use than anticipated and failed to meet expectations for sample delivery in the early Apollo missions. The manual interfaces and procedure for use were modified to reduce resistance of the regolith, which was denser than anticipated due to compaction when force was applied.

Fig. 6.6 Surveying Staff being field tested with Lunar Geological Exploration Camera (LGEC) (top left and right), with laser range finding reflector and spectrographic systems as described in the text [12]. (Courtesy of USGS)

For the manual drive tubes:

- the Diameter of the shallow drive tubes was increased from 2 to 4 cm for the J missions.
- the design of the end effector (bit) was redesigned for Apollo 12, and then again for the J missions

Fig. 6.7 Apollo Geology Tools. Top Proposed Desktop-Sized thin section and slicing machine from Microtek that never flew due to limitations in allowable equipment mass. Bottom: 2-cm diameter core tube attached to shorter style extension handle being driven into the regolith on Apollo 12. (Courtesy of NASA)

- More force (via hammering) was applied on later missions to force the tube into the regolith
- the follower, providing a restrain on the upper limit of the core, was replaced with a keeper plus a rod acting as a ram to increase compaction of the sample prior to removal

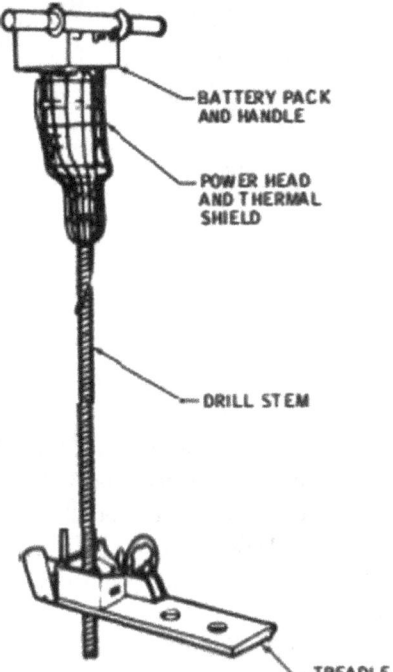

Fig. 6.8 Power Drill used on the Apollo J missions being set up by Astronaut Irwin on Apollo 15 (top) with drawing of major components below [11]. (Courtesy of NASA JSC)

Once the core tube was forced into the soil, the tube was turned horizontally, the core tube extracted vertically, and a cap applied to the bottom to prevent sample loss.

Operating the powered drill required the astronaut to grip its handles with his gloved hand and to apply constant downward pressure, once again proving more difficult than expected, just as for the drive tubes. The drill 'was screwed into the soil by external flutes' (rather than being held down by a treadle as originally intended) and was removed, with difficulty, 'by powered action at constant depth' to clear of cuttings.

- 6 core stems were used for Apollo 15 and 16, and 8 core stems for Apollo 17
- The design of the core stems (tubes) was modified after Apollo 15 [13, 14].
- The rotating spiral groove flutes on the tubes were designed to extract material.
- The fiberglass/boron joints between tubes, initially tapered and apparently causing the tube to jam in the bore hole, were replaced by threaded titanium inserts creating continuous spiral grooves.
- The core stems were also lengthened to decrease the number of joints.

The powered drills were also used to create holes for emplacement of heat flow probes.

References

1. Schaber, G. Chronology of Activities from Conception through the End of Project Apollo 1960-1973, USGS Astrogeology Branch, Open File Report 2005-1190, http://pubs.usgs.gov/of/2005/1190/, 2005.
2. Clark, P.E., Revolution in Field Science: Apollo Approach to Inaccessible Surface Exploration, Earth, Moon, and Planets, 106, 2–4, 133–157, http://www.springerlink.com/content/uk63t1n6426016l7/, 2010.
3. Jones, E. and K. Glover, Apollo Lunar Surface Journal, http://nasa.gov/history/alsj/. visited 10/1/24.
4. Bailey. N., G. Ulrich, and K. Edmonds, Apollo Applications Program Field Test 8 with Section on Task Analysis, USGS Technical Letter 26, 1967.
5. Clark, P.E., C.R. Weisbin, K.E. Shelton, J.H. Smith, J. Mozinski, W. Lincoln, A. Elfes, H. Hua, V. Adumitroaie, R. Silberg (2009b) Lunar surface mission productivities as a function of mission duration and weighted science objectives, Space 2009, AIAA–2009–6632–995.pdf. http://www.aiaa.org/agenda.cfm?dateget=all, 2009.

6. C.R. Weisbin, P.E. Clark, A. Elfes, J.H. Smith, J. Mrozinski, V. Adumitroaie, H. Hua, K. Shelton, W. Lincoln, R. Silberg, M. Lupisella Reconciling Scientific Aspirations and Engineering Constraints for a Lunar Mission via Hyperdimensional Interpolation, AIAA Space Volume 4, p. 2648–2652, 2010.
7. USGS Astrogeology Branch, Astroshare Videos, Apollo Astronaut Training Films (originally 16 mm and translated into mp4 format), edited by Marc Hunter, 1968-72.
8. Ables, P., Time and motions required to perform an active seismic experiment proposed for the first Apollo landing, USGS Technical Letter, Astrogeology 12, 1967.
9. Bland, D., R. Blevins, J. Olmsted, R. Koppa, Final Lunar Surface Procedures, Volumes 1 and 2, 1972.
10. Phinney, W., Science Training History of the Apollo Astronauts, NASA, SP-2015-626, 2015. https://ntrs.nasa.gov/api/citations/20190026783/downloads/20190026783.pdf. visited 10/1/24.
11. Allton, J., Catalog of Apollo Lunar Surface Geological Sampling Tools and Containers, JSC-23454/LESC-26676, http://www.lpi.usra.edu/lunar_resources/documents/LunarSampleToolCatalog.pdf, 1989.
12. Dodenhoff, K., Apollo Lunar Geological Exploration Camera (LGEC), 2006. https://history.nasa.gov/alsj/alsj-LGEC.html. visited 10/1/24
13. CollectSpace, Apollo 13 lunar drill design evolution. http://www.collectspace.com/ubb/Forum29/HTML/002106.html. visited 10/1/24
14. Carrier, D., Apollo 15 Drilling Problems, Apollo 15 Lunar Surface Journal, https://nasa.gov/history/alsj/a15/a15carrier.html, 2013.

7

Surface Deployed Experiments

7.1 Apollo Lunar Science Experiment Packages (Fig. 7.1)

Although the Apollo science activities were focused on using geological traverses to sample and visually document the lunar landscape, NASA also supported opportunities for individual scientists to propose experiments to be deployed by astronauts on the lunar surface [1–3]. In this way, the Apollo program foresightedly represented and served the needs of a broad cross-section of the scientific community. The program was announced in 1963, with the National Academy of Science Space Science Board identifying major areas for study 2 years later. NASA then selected the experiments and contracted with Bendix Corporation to build the packages in 1966 [3–5]. These experiments would provide more comprehensive knowledge of the physical nature of the lunar surface, near-surface, and interior as well as the surrounding environment, and its interactions with the surface. Table 7.1 lists experiments carried by each Apollo mission [2]. Table 7.2 summarizes the highest priority measurement objectives [3].

The challenging operational requirements for the deployed science packages, which Bendix met, included [3]:

- operation in the deployment environment for 1 year (increased to 2 years for Apollo 17)
- manual deployment of experiment packages and supporting subsystems within astronaut capabilities and established safety constraints

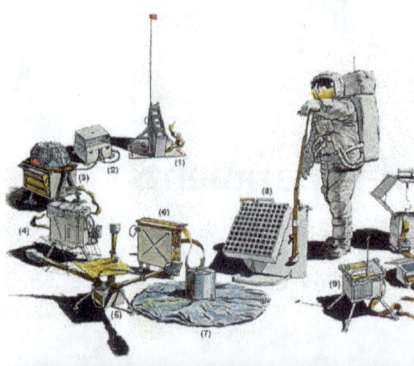

1) ASE Mortar Package Assembly
2) Heat Flow Experiment electronics box
3) Solar Wind Spectrometer
4) Suprathermal Ion Detector/Cold Cathode Ion Gauge
5) Lunar Surface Magnetometer
6) Charged Particle Lunar Environment
7) Passive Siesmic Experiment
8) Laser Ranging Retroreflector
9) Lunar Ejecta and Meteorites Experiment
10) Lunar Atmospheric Composition Experiment
11) Lunar Surface Gravimeter

Fig. 7.1 Top: Labeled drawings of many ALSEP instruments [2]. Bottom: ALSEP package deployed on Apollo 16 with Magnetometer in foreground and Astronaut Young approaching the central station in the background [2]. (Courtesy of NASA)

Table 7.1 Apollo surface deployed experiments [2]

Apollo Mission	Category	Experiment packages
11, 14, 15	Interior	LRRR Lunar Ranging Retro Reflector
11	Interior	EASEP PSEP Passive Seismic Experiment Package
12, 14, 15, 16	Interior	ALSEP PSE Passive Seismic Experiment
14, 16	Interior	ALSEP ASE Active Seismic Experiment
17	Interior	ALSEP LSPE Lunar Seismic Profiling Experiment
15, 16, 17	Interior	ALSEP HFE Heat Flow Experiment
17	Interior	ALSEP LSG Lunar Surface Gravimeter
17	Interior	TGE Portable Traverse Gravimeter Experiment
12, 15, 16	Interior	ALSEP LSM Lunar Surface Magnetometer
14, 16	Interior	LPM Lunar Portable Magnetometer
17	Regolith	SEP Surface Electrical Properties
11, 12, 14, 15	Regolith	EASEP/ALSEP LDD Lunar Dust Detector
17	Regolith	ALSEP LEAM Lunar Ejecta and Meteorites
17	Regolith	NP Neutron Probe
11, 12, 14, 15, 16	Environment	SWC Solar Wind Composition
12, 15	Environment	ALSEP SWS Solar Wind Experiment
12, 14, 15	Environment	ALSEP SIDE Suprathermal Ion Detecctor
12, 14, 15	Environment	ALSEP CCIG Cold Cathode Ion Gauge
14	Environment	ALSEP CPLEE Charged Particle Environmental Experiment
16, 17	Environment	CRD Cosmic Ray Detector
17	Environment	ALSEP LACE Lunar Atmosphere Composition Experiment

Table 7.2 ALSEP 'Observatory' measurement objectives [3]

Internal structure and composition of Moon
Heat flow from the lunar interior
Tectonic and meteorite impact processes, assessment of their importance in shaping lunar features
Near surface geological structures
Existence and nature of Moon's magnetic field
Interaction of solar plasma with Moon's magnetic field
Nature of and variations in lunar gravitational field
Characteristics of solar particles reaching lunar surface
Nature of Earth's magnetospheric tail
Nature of lunar atmosphere, composition of gases released by tectonic and impact processes
Precise orbit and libration pattern of the Moon

- capability to withstand natural and induced mission environments (launch, boost, descent to surface, surface temperature variations between -184 °C and 121 °C)
- full operation during lunar day and night

- communicating with Manned Space Flight network ground stations with downlink bit error rate of <10^{-4} and uplink bit error rate of <10^{-9}
- Compatibility with Lunar Module interfaces (internal volume 15 ft.^{3}, system weight constraints 300 pounds, stowed center-of-gravity constraints)
- deployability at lunar longitudes of ±45° and latitudes of ±25°
- deployability at sun angles of 7–25°
- Maintainability of system thermal control when all exposed surfaces are degraded by dust or ultraviolet radiation
- Capability of withstanding extended ground testing without damage

Thermal control for lunar surface packages is and continues to be extremely challenging. Bendix met operational requirements by utilizing a parabolic radiator reflector design [6] which continues to be used to this day, along with RadioThermal Heat Units (RHUs) and RadioThermal Generators (RTGs) as described below.

Experiment packages were stowed in the Scientific Equipment Bay and off-loaded through the use of booms and lanyards. The procedure was to deploy on level ground, for uncompromised thermal control, and 100 m to the west of the LM (but not in its shadow at sunset) to avoid dust and debris generated by the ascent vehicle. Apollo 11's Early Apollo Surface Experiments Package (EASEP) was powered by solar panels, but heated by a small RHU. Later Apollo missions carried the Lunar Surface Experiment Packages (ALSEP) powered by a Pu 238 based RTG called SNAP-27 (Fig. 7.2) [3–5]. The ALSEP packages had a central station to which each ALSEP element was connected by cable which distributed the power from the RTG (Fig. 7.2) and provided downlink to the Earth via an antenna gimballed to maintain Earthward pointing [3–5].

The Moon continues to be an especially challenging environment from both a thermal and radiation standpoint. Radiation protection is provided by the thickening the aluminum shell as needed. Bendix met thermal control requirements on ALSEP packages by utilizing a clever parabolic radiator reflector and sun shield design to deflect thermal emission from both the sun and the surrounding landscape in order to protect the equipment from overheating [6]. More than 50 years later, NASA's current requirements for lunar surface packages do not include night-time operation. Why? Today, Pu238, a strategic material, is limited in availability, exceedingly expensive, and thus beyond the means of small surface package program investigators. RTGs, which rely on thermocouples to convert heat to electricity, are very inefficient. The Multi-Mission (MM) RTG extensively used on Mars is about 6% efficient. The latest version, the General-Purpose Heat Source (GPHS) RTG is

7 Surface Deployed Experiments

Fig. 7.2 Top: Schematic of the ALSEP Central Station on the left which distributed power from the RTG on the right [2]. Bottom: Apollo 12 astronaut Bean offloading the ALSEP [2]. (Courtesy of NASA)

about the same efficiency but operates at a lower temperature. The GPHSRTG is also twice the size and mass of the MMRTG, and thus not very practical for small instrument packages. NASA announcements of opportunities to develop and deliver small packages to the Moon, through programs such as DALI and. CLPS, as interest in the Moon has grown over the last decade, has led to the development of far more efficient and compact components for removing or retaining heat [7]. Experiments, which can operate on at least a limited duty cycle during lunar night (without the use of RTGs) by taking advantage of these advances, are already being proposed today for small packages, to be discussed in more detail in Chap. 9.

7.2 Lunar Interior Experiments [8–13]

The astronauts deployed instruments designed to characterize the lunar interior on every mission. Experiments varied from mission to mission and included packages which provided seismic and heat flow, as well as gravity, and magnetic field measurements.

Lunar Ranging RetroReflector (Fig. 7.3) [14–16]: This experiment, deployed at the Apollo 11, 14, and 15 landing sites and aligned precisely to face the Earth, was designed to allow determination of the distance between the lunar surface and the Earth as a function of lunar orbital position. It consisted of an array of retroreflecting silica cubes that would reflect laser light pulses sent from a terrestrial source back to that source. The time elapsed between transmitting and receiving pulses could then be used to calculate the distance to within 8 cm. LRRR derived measurements did in fact show considerable 'wobbling' of the lunar surface due to tidal stresses as the Moon's distance and velocity vary as a function of position in its elliptical orbit. Although its surfaces experienced some solar degradation over time, laser technology continued to improve, and LRRR continued to provide useful measurements for decades. Unlike the other experiments, it did not require RF communication with the Earth to deliver data. RF communication with surface experiments was terminated in 1977.

Seismic Experiments (Figs. 7.4 and 7.5) [17–25]: Seismic experiments were deployed with the goal of providing detailed information on the Moon's internal activity and structure on all of the Apollo missions. These included passive seismic experiment packages (PSEP and PSE) and active seismic experiments (ASE and LSPE). These experiments operated on a stand-alone basis via solar panels, and thus did not take measurements during lunar night. Passive seismometers were deployed on Apollo 11 through 16, its original

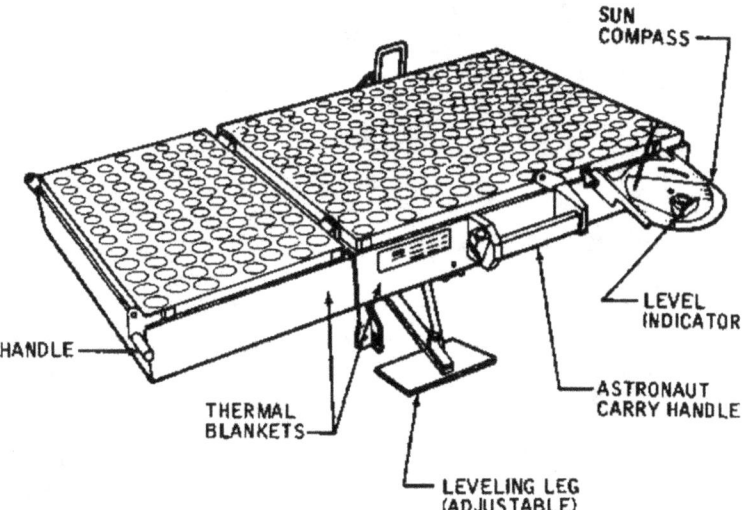

Fig. 7.3 Laser retroreflector deployed (above) and drawn with label parts (below) [2]. (Courtesy of NASA)

design modified to prevent the overheating that apparently occurred and shut down the Apollo 11 seismometer prematurely. Data were recorded continuously from Apollo 12, 14, 15, and 16 passive seismic stations until 1977. Active seismic experiments, deployed with geophones, were performed at Apollo 14 and 16, and the Lunar Seismic Profiling Experiment, which

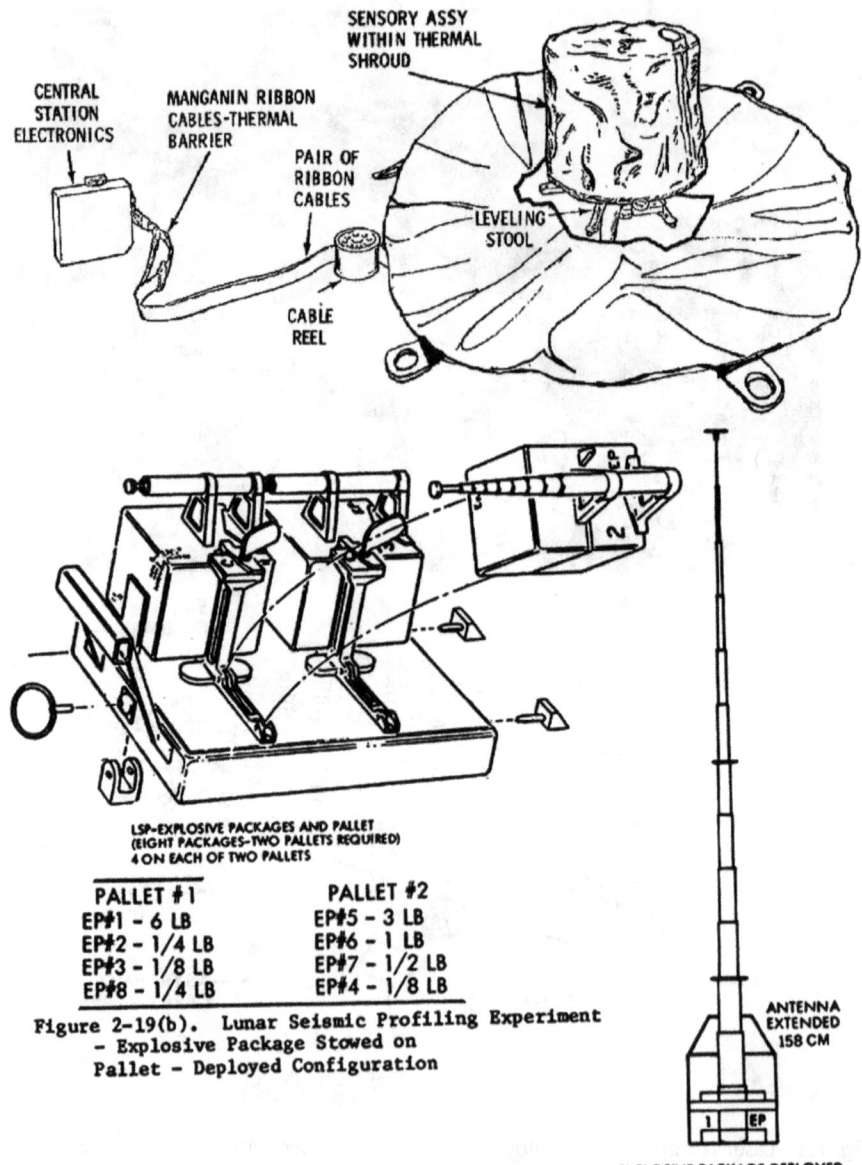

Figure 2-19(b). Lunar Seismic Profiling Experiment - Explosive Package Stowed on Pallet - Deployed Configuration

Fig. 7.4 Seismic Experiments. Passsive Seismometer (top) and active seismometer components with explosive packages (closeup on right) mounted on pallet to be deployed along a line of geophones [2]. (Courtesy of NASA)

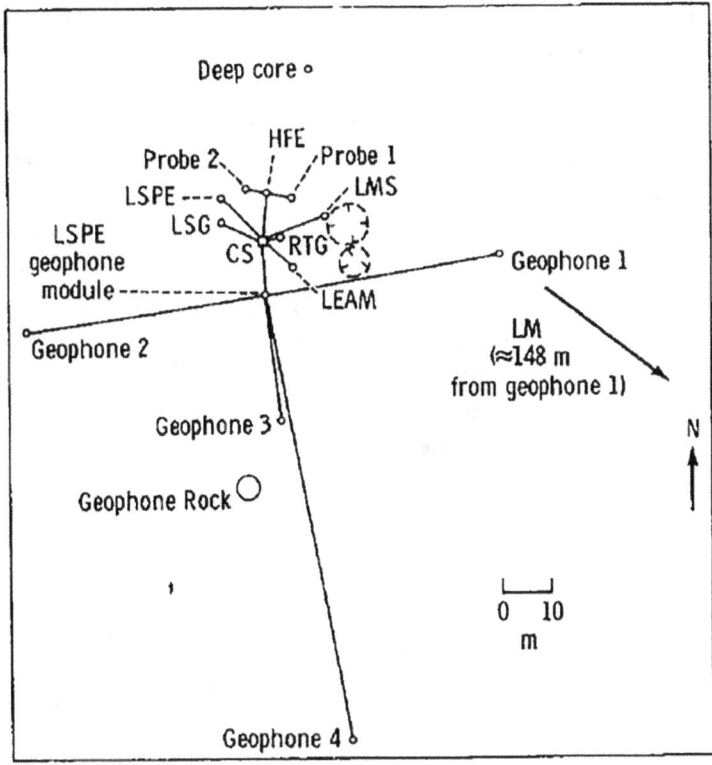

Key: CS = central station
HFE = heat flow experiment
LEAM = lunar ejecta and meteorites experiment
LMS = lunar mass spectrometer
LSG = lunar surface gravimeter
RTG = radioisotope thermoelectric generator

Fig. 7.5 Distribution of 4 geophones round the LSPE lunar seismic profiling experiment on Apollo 17 [2]. (Courtesy of NASA)

included a lunar surface gravimeter to provide additional seismic information, was performed at Apollo 17.

The passive seismometers were designed to measure small surface vibrations, including the most common from meteoritic bombardment, and less frequent interior activity (moonquakes), which turned out to be two orders of magnitude less likely than external bombardment. Interior activity increased when the Moon was at perilune and apolune positions. Rapid heating at sunrise also generated recordable vibrations. Planned human-generated impacts and explosions (e.g., Saturn V third stage and LM ascent stage impacts) were used to calibrate the sensors. The passive seismometer had both long-period displacement sensors (two horizontal and one vertical) and a short period vertical displacement sensor. Internal motors kept the instrument level to within a few second of arc. Deployment of these experiments made three-dimensional triangulation of internal events and understanding of the Moon's internal structure, possible. These measurements allowed the composition and size of the lunar interior (plagioclase rich crust, olivine and pyroxene rich mantle, <450 km radius iron and sulfur core) to be roughly constrained. Moonquakes appear to be mostly induced by lunar tidal stresses. Seismic waves are greatly attenuated in the upper crust by scattering from heavy fracturing due to meteorite bombardment, much less attenuated in the Moon's colder and dryer (than Earth's) mantle, and then more attenuated again at 1000 km, possibly indicating the presence of partially molten rock. Shock waves from spacecraft impact on the Moon caused long periods of vibration, increasing and then gradually decreasing in amplitude for periods of many minutes, called 'ringing', thought to be due to the extreme lack of moisture causing 'dampening' of the signal.

The active seismic experiment generated shocks that were measured by geophones from two types of charges. For the 'thumper' experiment, on Apollo 14 and 16, astronauts set up a straight line of geophones and used a staff with 19 explosive charges (thumper) to fire shotgun-like charges at evenly spaced intervals between them. Charge detonation then provided measurements from which local structure could be deduced while the astronauts were still on the surface. On Apollo 16, the astronauts also set up a mortar package armed with four rocket-launched grenades set for progressively greater distances, and with the capability for determining horizontal distance between the launch and impact sites. The astronauts aligned the rockets with a line of geophones and armed them for firing. Mortar shells were launched by radio control after their departure. The size of the explosive charges increased with distance, and allowed geophones to detect structural variations within 300 m of the surface. Another active seismic experiment, the Seismic Profiling Experiment, was deployed only at Apollo 17 (Fig. 7.5). Four geophones were deployed (three in an equilateral triangle with 100-m sides, and one roughly in the center of the triangle) near the landing site, instead of in a line (see comparison figure). Eight charges were distributed along traverse routes from 0.1 to 3 km away. The SPE was designed to provide information from the surface to greater depth, kilometers as opposed to hundreds of meters for the ASEs. The explosive charges, identical except for size of charge, were remotely detonated at different times after the astronauts' departures. After the detonations, the experiment became a passive seismometer.

All of the seismic experiments, both passive and active, indicated much lower than terrestrial velocities in the upper few hundred meters below the surface, consistent with highly fractured material anticipated for a surface subjected to constant bombardment, as opposed to solid, rock.

Heat Flow Experiment (Fig. 7.6) [26, 27]: This experiment was a subsurface probe designed to measure the rate of heat loss from the interior, as well as surface temperature and thermal properties of the regolith. Heat is generated in the interior due to natural decay of radioactive elements, and then dissipates toward the cooler surface and ultimately to space. The difference in temperature as a function of depth along a vertical probe will indicate the rate of loss, and from those measurements the abundance of remaining radioactive nuclides, and the current temperature, degree of melting to generate magma, and level of tectonic activity in the interior, can be inferred.

According to heat flow experiment measurements the temperature varied between a minimum (night time) (~75 K) and maximum (daytime) (~350 K) at the surface at the Apollo landing sites, but remained at a constant ~250 K at 1.5 m depth. These results indicated (1) that the lunar regolith is an excellent insulator, and (2) that the Moon, with an average heat flow of ~20 mW/m^2, has been far less geologically active than the Earth (average heat flow 87 mW/m^2). Heat Flow observations are consistent with lunar sample and photogeological data. As the Apollo measurements were taken near boundaries between the highlands and the (most recently geologically active) maria, they are thought to represent heat flow values that are ten to twenty percent higher than the global average.

The experiment required the use of the power drill, available on the J missions (Apollo 15, 16, and 17). The Apollo 16 HFE failed due to a broken cable. The drill was used to create not only one but two 1-in. diameter, 2 m-deep holes into the regolith for two sets of temperature sensor measurements. The second set verified the results of the first. To measure differential temperature within each hole, several platinum resistance thermometers were placed in the lower part of the hole, and several thermocouples in the upper part of the hole. Monitoring over several diurnal cycles allowed the effect of near surface daytime heating to be modeled and removed.

Only one of the two Apollo 15 heat flow probes was successfully emplaced to full depth. There was evidently an obstruction in the drill string in Hole No. 2, believed to be due to separation of two stems that occurred when the astronauts tried to overcome binding of the stems in the hole. A re-design of the stem joins eliminated the problem of binding on Apollo 16 and 17.

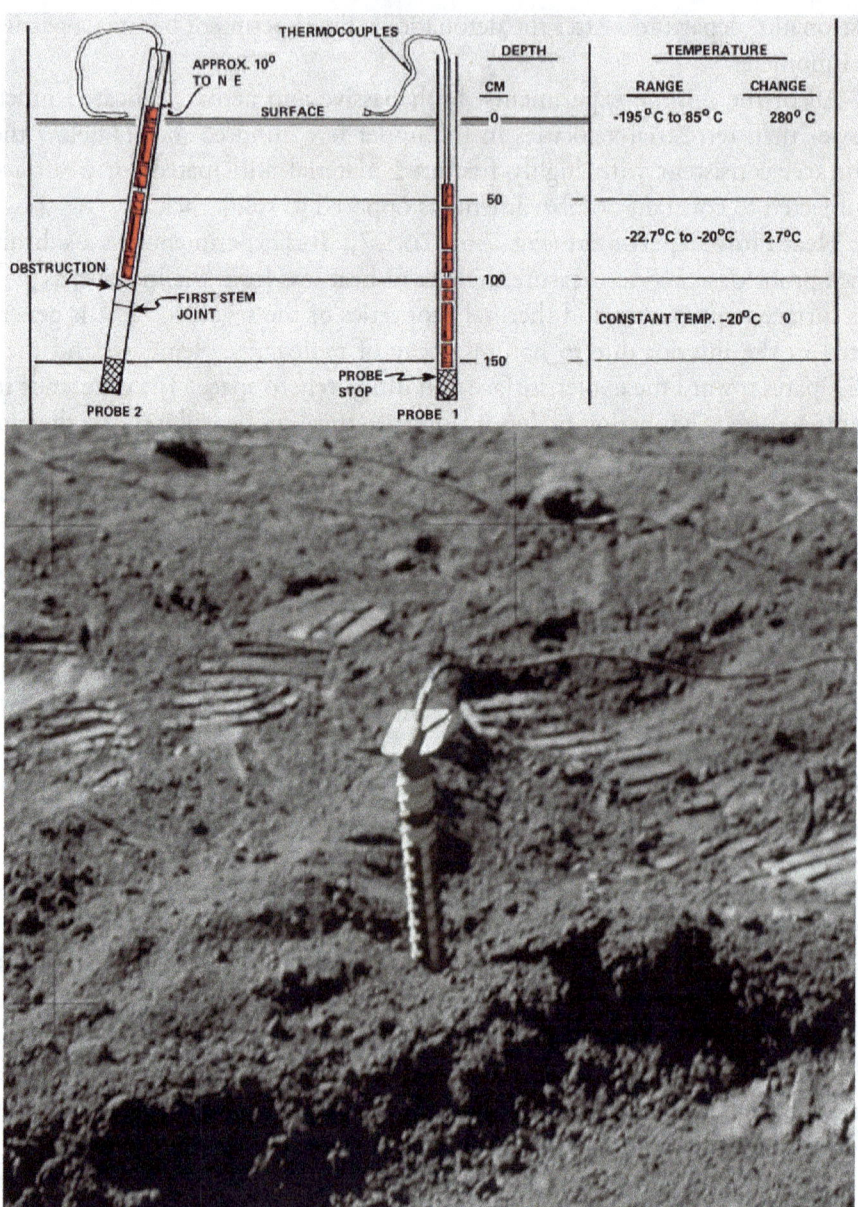

Fig. 7.6 Heat Flow Experiment Schematic of proper deployment finally accomplished on Apollo 17 (top). Below, first probe successfully deployed in hole made by power drill on Apollo 17 [2]. (Courtesy of NASA)

Unfortunately, no data was returned from Apollo 16 after the HFE ribbon cable was accidentally sheared off at the base of the Central Station. Good data was obtained from the Apollo 17 emplacement, which confirmed the Apollo 15 results.

Gravimeters (Fig. 7.7 and 7.8) [28, 29]: Two types of gravimeters were delivered to the lunar surface on Apollo 17, the lunar surface gravimeter (LSG) and the portable traverse gravimeter (TGE).

The TGE utilized a vibrating string accelerometer: by measuring the change in vibration due to weight induced by gravity (gravitational acceleration) on two wires with current passing through in the presence of a magnetic field. Measurements were taken at the landing site and at stops along traverses. Variations in gravitational acceleration indicated changes in underlying mass and structure. These measurements, when combined with topography measurements derived from orbital stereo photography, would provide important constraints and confirmation on the interpretation of seismic data, indicating the presence of a basalt layer 1–1.4 km thick underlying the landing site.

The LSG, essentially a spring balance acting as a one axis seismometer, was designed to measure temporal variations in lunar gravity as a function of lunar orbital position and tidal distortions. Due to a mathematical error made in configuring the instrument (for 1/6 g instead of 1 g), insufficient mass was available to null the beam, and thus it was unusable on the Moon.

Magnetometers (Fig. 7.9) [30–33]: Two experiments with three-axis fluxgate magnetometers were used to measure magnetic field intensity and variations on the lunar surface. The Lunar Surface Magnetometer (LSM) measured temporal variations in the magnetic field on the lunar surface at several landing sites. The Lunar Portable Magnetometer (LPM) made measurements of spatial variations in the magnetic field along traverses induced by variations in magnetic properties of underlying deposits, analogous to prospecting for buried ore bodies on Earth. The Lunar Orbiters had already measured a steady magnetic field component of about 10 nT (nanotesla), as well as time variable component induced by waves propagated from the sun (the Interplanetary Magnetic Field or IMF). The lack of a global magnetic field indicated no internal dynamo. An orbital electron analyzer, flown on Apollo 15 and 16, detected backscattered electrons which were correlated with the presence of local magnetic fields associated with impact basins and generated a via shock magnetization large meteorite impact events.

The LSM was deployed at Apollo 14, 15, 16, and 17 landing sites, measured a steady magnetic field of 36 nT, the subsurface resistivity and temperature dependent induced magnetic field, with two variable superimposed fields: the interplanetary magnetic field (IMF) and the Earth's geomagnetic

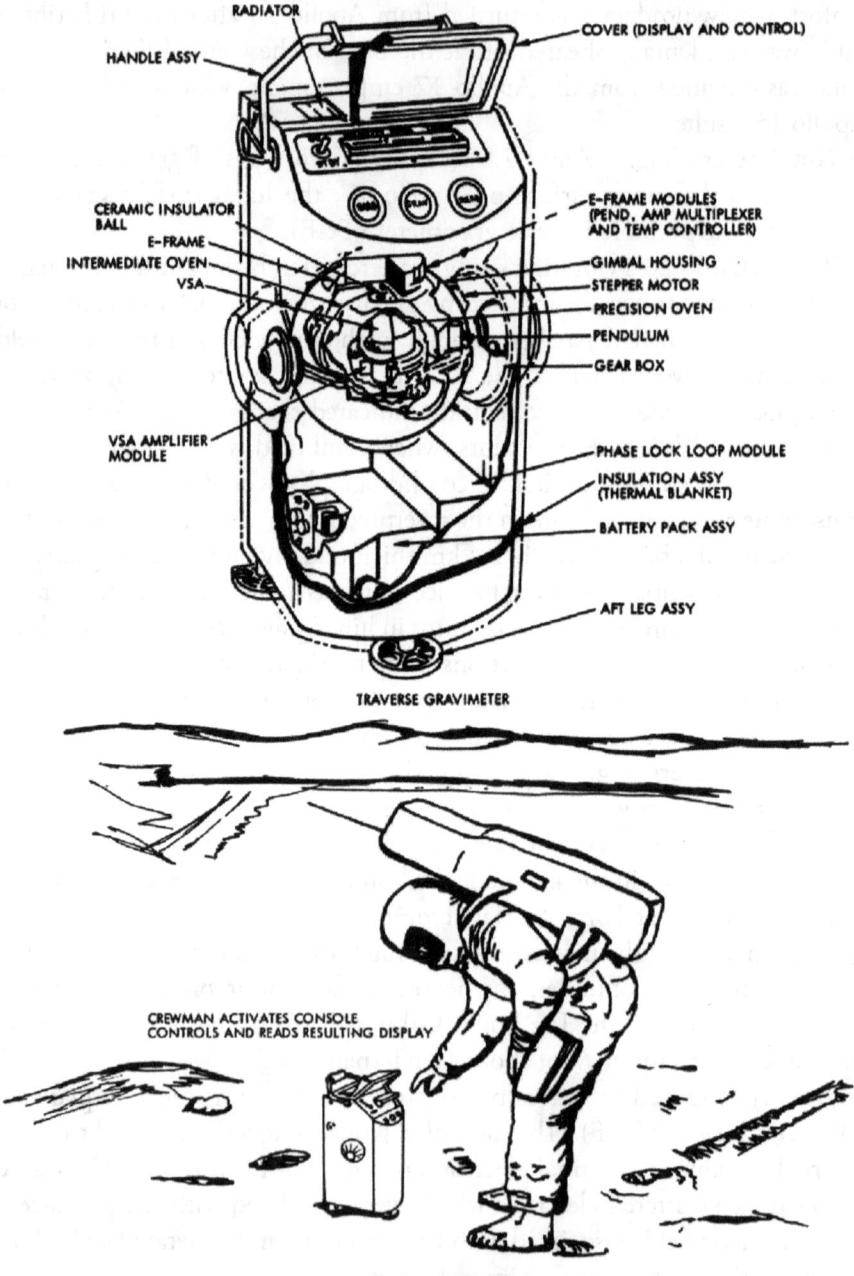

Fig. 7.7 Drawing of the portable Traverse Gravimeter Experiment (TGE) (above) deployed on the lunar surface (below) [2]. (Courtesy of NASA)

Fig. 7.8 Traverse Gravimeter Experiment (blue gray box) mounted on the back of the Apollo 17 rover. Astronaut Cernan deployed and set up the TGE, took measurements, and remounted it at each station. He took this shot to document the 'replacement fender' made from the back of the manual [2]. (Courtesy of NASA)

tail when the Moon passed through. The non-varying component of the magnetic field is thought to be a remnant field remaining (and gradually decreasing) from a time very early in its history when the Moon had an internal dynamo and thus a global magnetic field. The magnetometer also showed a response to diurnal variations in the solar wind. Thus, these measurements, as well as seismic and gravimetry measurements, contributed and confirmed our understanding of the Moon's interior structure. The 50-km crust is heavily fractured, multi-layered with a secondary boundary at 20 km. The 500-km upper mantle is homogeneous, and dominated by olivine and olivine-pyroxene rich rocks. Below is an iron-rich interior, and possibly a molten core.

LPM measurements were taken at locations along traverses on Apollo 14 and 16 and made with magnetometer mounted on a tripod connected to the rover with a ribbon cable, far enough (15 m) away from the vehicle to prevent

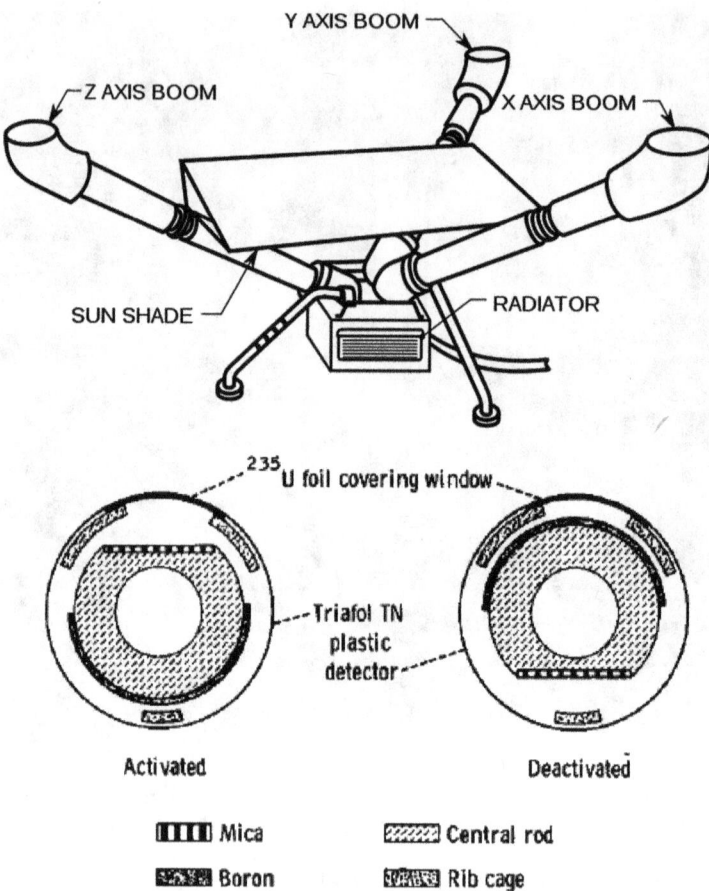

Fig. 7.9 Top: Schematic of the Magnetometer. The magnetometer deployed on the surface can be seen in Fig. 7.1 [2]. (Courtesy of NASA). Bottom: Lunar Neutron Probe schematic cross-section illustrating activated (boron targets and mica detectors on central rod face plastic detectors and uranium targets on rib cage) mode on left and deactivate (targets and detectors do not face each other on right [13, 34]. (Courtesy of NASA Apollo 17 preliminary science report)

magnetic interference. Each of the three digital panels at the rover gave readouts of field intensity for each axis.

Based on the lunar magnetic field strength being three orders of magnitude smaller than the Earth's (50,000 nT), range of the LPM was set to ±128 nT. However, measurements (43 nT and 103 nT on Apollo 14) and the differences between them over a short (hundreds of meters) differences, were far larger than expected, particularly when previous magnetic field measurements of the Moon from orbit indicated an average field of 10–12 nT. A similar result was obtained on Apollo 16, confirming the likely cause as the natural magnetization of lunar rocks. This process occurs on Earth when iron ore, most commonly magnetite, is permanently magnetized through a lightning strike, forming lodestone. Meteorite impacts are thought to play a similar role on the Moon.

Neutron Probe NPE (Fig. 7.9) [34]: The Neutron Probe Experiment (NPE), flown on Apollo 17, was one of several experiments used to provide information on the character of the lunar subsurface. In this case, the variation in radioactivity with depth would be used to determine the extent, or depth, to which gardening, or turnover of the regolith resulting from constant meteoritic bombardment, had occurred. The probe consisted of a 2.4-m rod with several sensors placed at different depths along the rod. The rod was inserted into the hole remaining after a deep drill core sample had been collected during the first EVA and removed (and turned off) at the end of the third EVA. Cosmic-ray protons may strike nuclei of atoms at and below the surface, generating neutrons which may interact with other atoms to generate radioactive isotopes. The depth to which increased radioactivity can be measured is used to determine the depth to which gardening occurs. Sensors consisted of cellulose triacetate plastic used to detect alpha particles generated by boron targets when neutron capture occurred, and mica detectors used to detect fission fragments in uranium 235 targets, from which the rate of neutron generation and resulting enhanced radioactivity could be derived. In order to derive that information, it is necessary to incorporate (1) the isotopic composition and background level of radioactivity known from analysis of lunar samples; (2) the radioactive decay rate for isotopes known from extensive laboratory experimental data; and (3) the rate at which radioactive elements are created by neutron bombardment known from measurements below the level of gardening with lunar neutron probe data collected in this experiment. Mixing to a smaller depth, resulting from far more frequent smaller impacts, occurs much more readily than mixing to greater depths, requiring the far fewer larger impacts. Measurements made by this experiment confirm, along with data from other sources as well, that mixing to a depth of 1 cm occurs within a million years, whereas mixing to a depth of 1 m requires 1 billion years.

7.3 Lunar Environment Experiments [8–13]

Lunar Dust Detector LDD (Fig. 7.10) [35, 36]: Flown on Apollo 11, 12, 14. And 15, the Lunar Dust Detector consisted of three photocells mounted on the central station used to detect changes in solar illumination by measuring

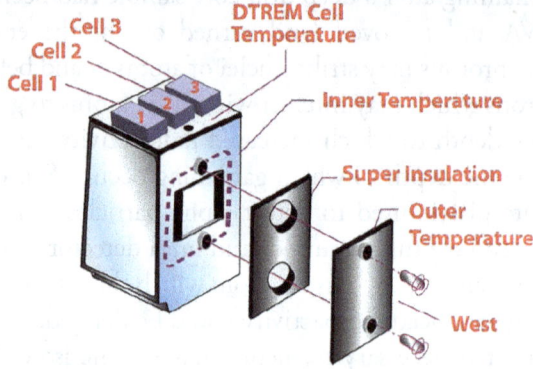

Fig. 7.10 Dust Detectors. Lunar Dust Detector (LDD) deployed on the lunar surface (top left) and Lunar Ejecta And Meteorites (LEAM) in the lab (top right) [2]. Courtesy of NASA. Bottom: Schematic of the LDD dust detector. (Courtesy of NASA/NSSDC)

power output and temperature from the cells. Removing the effect of normal diurnal changes, such changes could be induced by an accumulated dust layer or high-energy radiation damage to the cells. Changes seen were primarily due to gradual degradation due to radiation damage, with little noticeable change resulting from dust accumulation. Although LADEE found virtually no evidence for levitated dust at altitudes above 3 km, analysis of LDD data indicated a higher rate of dust accumulation at the sunrises immediately following the departure and creation of a dust plume by the ascent vehicle. O'Brien and coworkers interpreted the data to indicate that the disrupted dust particles, freer from cohesive forces, were subject to more effective charging and mobilization by solar photons at the next sunrise [36].

Lunar Ejecta and Meteorites LEAM (Fig. 7.10) [37]: Flown on Apollo 17, LEAM was designed to utilize three motion sensor plates oriented orthogonally to measure the frequency, velocity, and direction of small particles striking the Moon and their secondary ejecta. The frequency of incoming dust particles proved so small, that the experiment primarily detected the horizontal transport of fine, slow-moving dust grains at local sunrise. This result was good news from the standpoint of astronaut and equipment survival on the lunar surface.

Surface Electrical Properties SEP (Fig. 7.11) [38]: This experiment, flown on Apollo 17, involved transmitting continuous wave electrical signals from the landing site through the regolith and receiving them on the rover at several locations along its traverses, to determine the electrical properties (transmission, absorption, reflection, overall conductivity) of the regolith. Any moisture present would have resulted in changes in electrical conductivity by several orders of magnitude. This did not happen, proving that the regolith is extremely dry, consistent with lunar rock composition. The frequencies used were selected (1, 2.4, 4, 8.1, 16, and 32.1 MHz) to provide the basis for a geological model of the lunar subsurface from depths of a few meters to a few kilometers. The number and size of scattering bodies as a function of depth were derived from the detection of continuous wave signals transmitted at these frequencies and received by the rover. To set up the experiment, the astronauts transported the transmitter 100 m form the lander, and then deployed the multi-frequency transmitter antenna and wide-band orthogonal receiver antenna, and attached the receiver/recorder to the rover. Number of wheel-turns and difference between left and right wheel turns established the distance and direction of the rover relative to the lander. The recorder was then detached and returned with the astronauts. The SEP experiment also provided ground truth for the Bistatic Radar and Lunar Sounder experiments.

Cosmic Ray Detector CRD (Fig. 7.12) [39–41]: Cosmic rays are particles that have extremely large energies and very high velocities approaching the speed of light. They are mostly protons and alpha particles (helium nuclei) with a small percentage consisting of the nuclei of heavier elements. The cosmic rays seem to arrive from all directions and, although their origin is not yet known with absolute certainty, the more energetic ones appear to come from outside of our solar system. In addition to cosmic rays, the CRD equipment detected lower energy solar wind particles. The Cosmic Ray Detector (CRD) flown on Apollo 16, was used to detect the extremely energetic, and high velocity particles, mostly protons and alpha particles, coming from beyond the solar system, as well as much lower energy solar wind particles, by directly measuring their tracks. Collector panels, mounted in the LM, were an array of plates consisting of two sets of 31 sheets of 0.025 cm Lexan covered by perforated aluminized Teflon, four sheets of 0.2 cm

Fig. 7.11 Solar Electrical Properties (SEP) Experiment being deployed on the lunar surface by Apollo 17 astronaut Schmitt (top) and a closeup with labels before deployed [2]. (Courtesy of NASA)

Kodacell cellulose triacetate sheets overlaid by ten 5-μm Lexan sheets on the upper half, plus several plates of special materials. Panels were returned to Earth for analysis. The nature of a track reveals the type of particle and its direction of travel.

Fig. 7.12 Cosmic Ray Detector (CRD) (left) and Solar Wind Composition Experiment (SWC) (right) as described in the text [2]. (Courtesy of NASA)

Major findings were that (1) the heavier (iron) nuclei were enriched relative to lighter nuclei in the lower (solar origin) energy region, particularly during solar flares; (2) in the higher energy region, nuclei abundances were comparable to those previously observed for galactic cosmic rays caused by spallation reactions in interstellar space.

Solar Wind Composition Experiment SWC (Fig. 7.12) [42–46]: Flown on Apollo 11, 12, 14, 15, and 16 missions, The SWC was designed to measure the elemental and isotopic composition of the solar wind noble gasses (He, Ne, Ar). The experiment consisted of a single 0.5 mm thick aluminum sheet that trapped individual gas molecules to a depth of several hundred atomic layers, but allowed cosmic rays to pass through. Astronauts deployed the sheets upon arrival and detached them for return to Earth for analysis at the end of the last EVA. He flux was found to be correlated with the level of solar activity (e.g., relatively lower on Apollo 12 and higher on Apollo 15). The noble gas isotope ratios were found to vary relatively little between the Apollo missions.

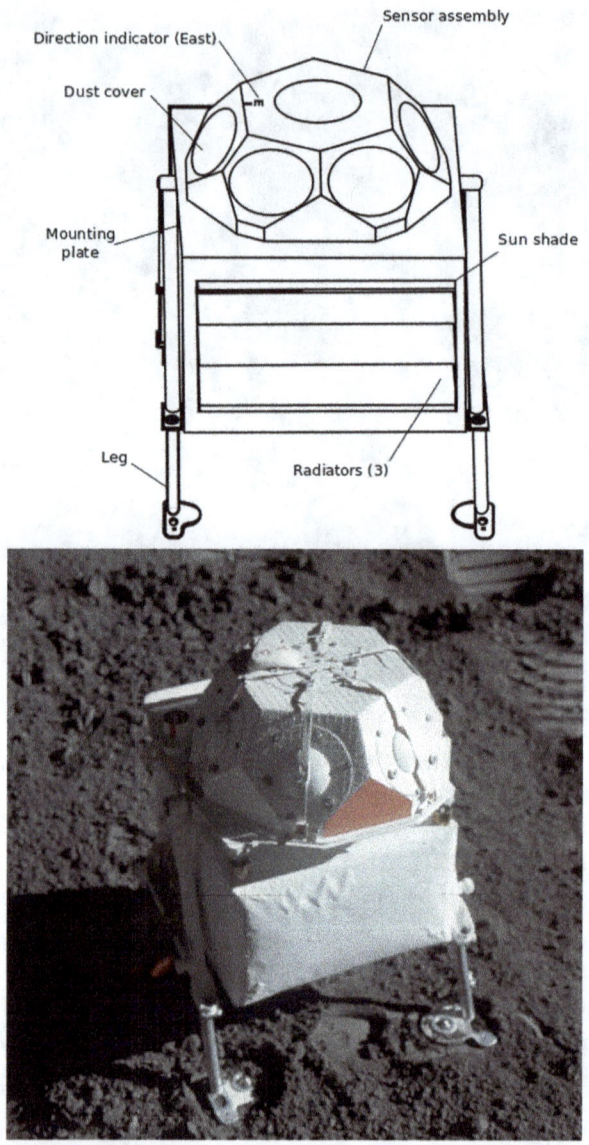

Fig. 7.13 Solar Wind Spectrometer (SWS) schematic (top) and after deployment on Apollo 15 (bottom) [2]. (Courtesy of NASA)

Solar Wind Spectrometer (Fig. 7.13) [47, 48]: The solar wind spectrometer, deployed on the lunar surface as part of the Apollo 12 and 15 ALSEP packages, was designed to measure the properties of protons and electrons at the lunar surface and their interactions with the Moon's surface and interior, as well as the interactions of the Earth's magnetotail with the Moon's surface and interior. The solar wind consists of matter constantly ejected from the sun stretching out to the edge of the solar system (and comprising the heliosphere). The experiment was designed to measure the energy, density, and direction of particles intercepted by the lunar surface. The SWS consisted of seven modulated Faraday cups opened toward different, but slightly overlapping, portions of the lunar sky. The instrument observed the directional intensities of the electron (6–1330 eV) and positive ion (18–9780 eV) components of the solar wind and the Earth's magnetotail plasma as the Moon passed through it monthly. Generally, the solar wind was observed to behave the same way at the lunar surface as it did in free space except that it was disturbed (no plasma measured) within the magnetotail. Surprisingly, the instrument continued to register encounters with charged particles for several minutes after the sun had set, indicating particles were traveling just beyond the terminator. The SWS also detected the oxygen gas cloud generated by the explosion of the oxygen gas tank on the Apollo 13 Service and Command Module. The explosion was also detected visually from a telescope set up on the roof of the mission control building at JSC.

Charged Particle Lunar Environment Experiment CPLEE (Fig. 7.14) [49]: The Charged Particle Lunar Environment Experiment (CPLEE), flown on Apollo 14 only, was designed to measure the ambient fluxes of charged particles, protons and electrons, in the 40–70,000 eV range, at the lunar surface. The objective was to understand the distribution of these particles in the Earth/Moon system, their relationship to the solar wind and their impact on Earth auroras, the van Allen radiation belt, magnetosphere (particularly the magnetotail) and magnetic field, as well as their impact as the solar wind 'shock front' on the lunar environment. The experiment consisted of 2 spectrometer packages oriented in two different directions for minimum exposure to the sun's ecliptic path. Each package had 6 particle detectors, which included 5 C-shaped channeltron photon-multipliers made of 1 mm diameter and 20 cm long glass capillary tubes and 1 helical funneltron photon-multiplier. Entering particles were subjected to varying voltages resulting in separation on the basis of charge and energy. Observations indicated the presence of low energy photo-electrons during the day, rapidly fluctuating low energy electrons in the magnetosheath and magnetotail, large rapid (within 10 s) changes in solar wind flux, regardless of the Moon's orbital position. Unexpectedly, electrons with terrestrial aurore band energies were detected in the Earth's magnetotail. CPLEE also detected plasma clouds from the Apollo 14 ascent vehicle impact.

Lunar Atmospheric Composition Experiment LACE (Fig. 7.15) [50]: The Lunar Atmosphere Composition Experiment, flown on Apollo 17, was

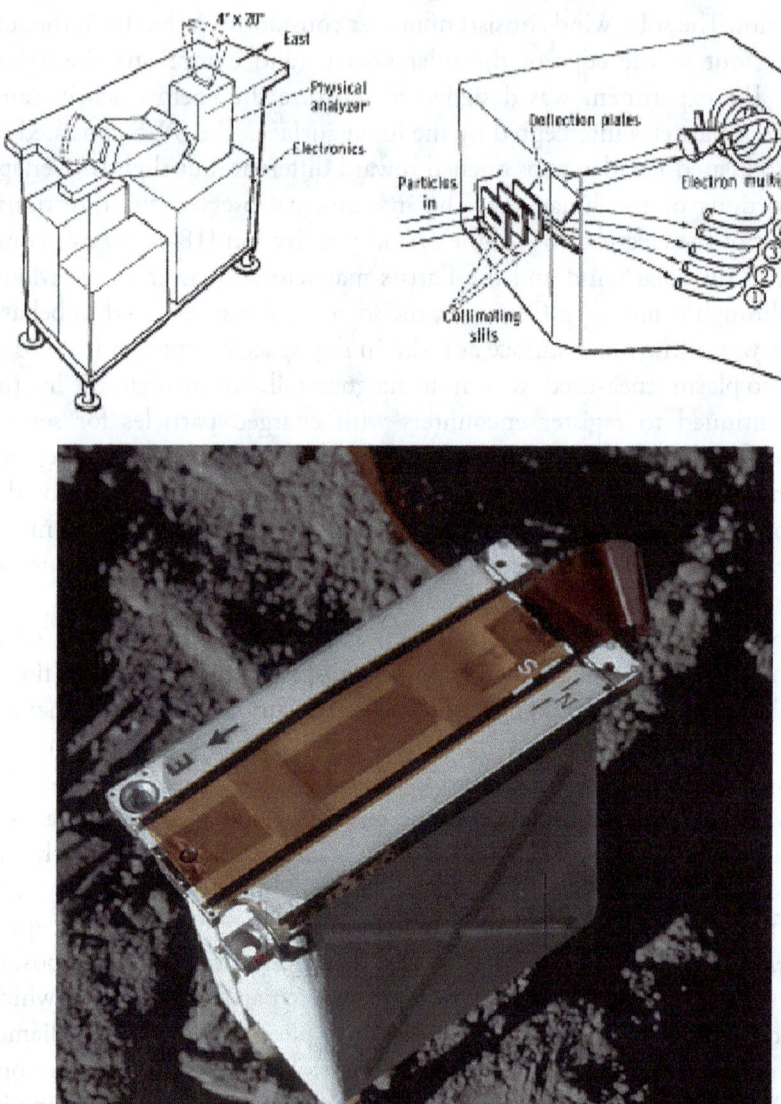

Fig. 7.14 Charged Particle Lunar Environment Experiment (CPLEE). Interior schematic of charged particle analyzer and sketch showing fields of view and look direction (top) and deployment on Apollo 14 [2]. (Courtesy of NASA)

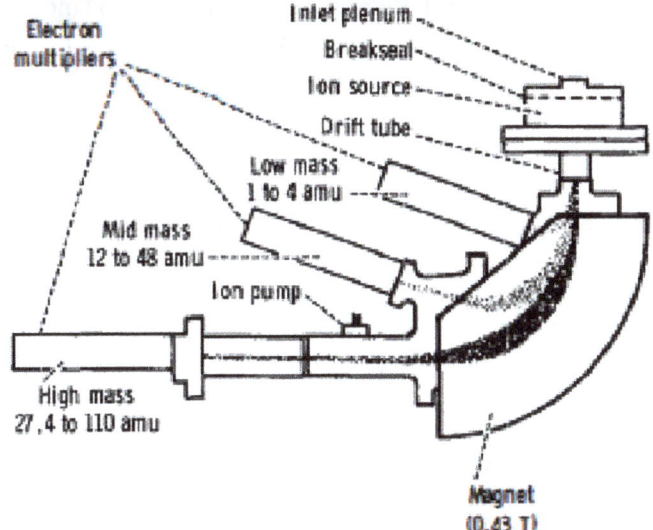

Fig. 7.15 Lunar Atmospheric Composition Experiment (LACE). Deployment on Apollo 17 (top) and schematic of magnetic deflection mass spectrometer (bottom) [2]. (Courtesy of NASA)

designed to provide direct measurements of the lunar atmospheric constituents. As a result of its low gravity, the Moon's atmosphere, actually a surface-bound exosphere with no interactions between individual constituents, is easily lost to space. For example, H_2 and He escape within a few hours. Heavier atoms are ultimately ionized and carried off by the solar wind within a few months. The exosphere is thus maintained by temporary capture from continuous source(s), most likely the solar wind, but also potentially including incoming meteoroids and comets and interior outgassing (especially He^4 and Ar^{40}).

LACE consists of a miniature magnetic deflection mass spectrometer, where gases are ionized in an electron beam, collimated, and sent through a magnetic analyzer. The ion mass number is correlated with the ion accelerating voltage, varied in stepwise fashion. Three magnetic field regions (three slits and three trajectories) are swept simultaneously, allowing a wide mass range to be explored with a relatively narrow voltage excursion. LACE was deployed with aperture zenith facing to intercept downward flux of atmospheric constituents which included primarily H_2, He, and Neon in equal abundance. Minor constituents observed included atomic hydrogen, nitrogen, oxygen, methane, carbon dioxide, ammonia, and water, chlorine, and hydrochloric acid, many of which were undoubtedly contaminants from landing site activities and declined with the passage of time. The experiment represented an advance over the measurement of total atmospheric pressure with implications for composition by the Cold Cathode Ion Gauge Experiment flown on earlier missions. Ar^{40}, the daughter product of K^{40} decay, was observed to increase during times of seismic activity, indicating outgassing from newly formed fractures. He abundance increased by a factor of 20 at midnight, indicating two things: its source is the solar wind, and it does not freeze out during lunar night. Argon decreases during lunar night and rises again at sunrise, as expected for Argon which has an interior source and freezes out. The mass spectrometer mass 19 peak, assumed to be Neon, has been seen to decrease during lunar night on some occasions, and to rise during lunar night on others. Neon should not be a condensable gas, so a decrease would not be expected. Is the mass 19 peak Fluorine and the variability due to outgassing of a contaminant in the instrument (lace, prelim sci report)? Most contaminants freeze out during lunar night and rise rapidly at sunrise. Except for Ne, all measurements are generally in good agreement with predictions.

Cold Cathode Ion Gauge CCIG and Suprathermal Ion Detector and Mass Analyzer Detector SIDE (Fig. 7.16) [51–56]: The CCIG and SIDE experiments, deployed in tandem and connected via a short electrical cable, on the Apollo 12, 14, and 15 missions, were designed to characterize the lunar

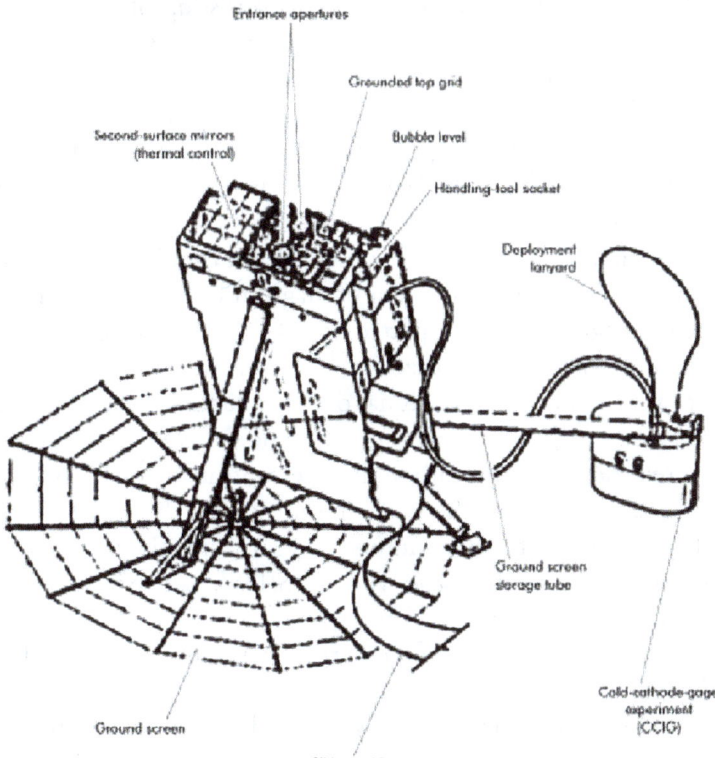

Fig. 7.16 Cold Cathode Ion Gauge (CCIG) and Suprathermal Ion and Mass Analyzer Detector SIDE for Apollo 15 mission reflecting significant redesign from earlier missions as described in text. CCIG required input from SIDE to operate [2]. (Courtesy of NASA)

atmosphere (exosphere). Difficulties in deployment of the two instruments side by side led to significant redesign for the Apollo 15 version of the instruments as shown below. Characterization involved measuring the composition and pressure of tenuous gases produced from outgassing or solar wind interactions with the surface, as well as the physical and energetic properties of positively or negatively charged particles in the eV to KeV (low to mid) energy range generated either by ionization of exospheric gases in situ or generated outside of the local lunar environment. Most of these ions would originate from Hydrogen and Helium consistent with solar wind composition. The experiments were designed to determine the extent to which an additional component, from lunar interior outgassing or impact events, is contributing to the exosphere.

The CCIG sensor is a coaxial electrode, consisting of a central cathode spool surrounded by a cylindrical anode surrounded, with a magnetic field applied to the along the axis. Voltage is applied to the anode to develop a self-sustaining electrical discharge. The current measured at the cathode is correlated with gas density. Deriving compositional measurements from this data requires input from the SIDE instrument. Early observations showed pressures and pressure variations primarily influenced by outgassing from human (lander) activities. Colder nighttime observations, thought to represent natural ambient conditions, were an order of magnitude lower, indicating a much lower rate of outgassing than expected. The CCIG measured the pressure of the lunar atmosphere at the surface at 10^{-12} torr, and total gas abundances ranging from $1 \times 10^7/cm^3$ during lunar day representing total gases to $2 \times 10^5/cm^3$ during lunar night representing non-condensable gases only. (Note: it also measured the water and carbon dioxide continually outgassing from the astronauts' space suits when they were in the vicinity.)

The SIDE measured the flux, number, density, directionality, velocity, and relative energy of the positive ions. The two SIDE positive ion detectors were oriented in the ecliptic plane 15° from the lunar local meridian, translating to 38° west, 2° west, and 19° east of the Earth for Apollo 12, 14, and 15 respectively. Thus, flows in the magnetosheath and events of several types related to the moon and magnetotail were observed, but detectors never looked directly at the sun (solar wind origin). The first, a mass analyzer, consists of velocity combined with electrostatic energy per unit charge filters on an ion counter, allowing determination of mass per unit charge. The second has the energy per unit charge filter only on the ion counter. Thus the instrument provided mass per unit charge measurements as a function of energy. A wire ground screen deployed beneath the instrument compensates for potentially large and varying lunar surface potential. Natural flux-generating events included partial eclipses, solar flares, geomagnetic storms. Recurring cycle events included sunrise and sunset (generating mid-energy particle clouds), night time magnetopause, magnetosheath, and bowshock crossings (generating energetic ions). Flux variations were correlated with diurnal variation in lunar surface temperature. Apparently, impact-generated seismic events influenced the direction of ion clouds, possibly corresponding to the release of subsurface water vapor. Human-generated events were also observed, including the change in flow associated with the generation of the exhaust gas cloud during lunar liftoff and the generation of an impact-gas cloud at the impact of the lunar ascent stage. Data from the experiments were used to determine the potential of the lunar surface and its effective plasma screening length, as well as to measure the magnetotail ion fluxes and ion mass spectra. Data were consistent with SWS measurements.

7.4 Major Results from Experiments

Samples collected, measurements made, and experiments run on the lunar surface provided the basis for major breakthroughs in our understanding of the origin and ongoing processes of the Moon, the Earth Moon system, and the solar system. Some of these results are as follows:

The Moon is a differentiated body, which has evolved to form a crust and internal structures over time, and not a primordial object. The highlands formed about 4.5 billion years ago from flotation of low density felspar rich rocks in a 'magma ocean', a completely molten surface layer, which formed the lunar crust. This was followed by a period of extensive bombardment by large impactors, creating large basins that were later filled by basalts during the period of extensive mare volcanism between 3 and 4 billion years ago.

The three major lunar rock types observed on the moon are igneous in origin, resulting from high temperature processes involving little or no water. These are the light, felsic anorthosites associated with the oldest and most ubiquitous terrain, the highlands; dark, mafic mare basalts, associated with flood plain volcanism which filled impact basins after the highland crust was formed; and breccias, resulting from melting and mixing occurring during the bombardment process.

Based on oxygen isotope ratios measured in lunar samples, the Moon and Earth have a common ancestor; however, the Moon is relatively depleted in volatiles and iron. Geochemical dating of lunar samples has established absolute ages for lunar rocks and the impact or volcano-tectonic events with which they are associated. The oldest rocks are as old as the solar system, 4.6 billion years, far older than rocks found on Earth, where the oldest, stratigraphically exposed rocks are less than 4 billion years old due to far more extensive volcano-tectonic activity and extensive resurfacing.

The Moon's surface thus acts as a 'time machine', revealing processes that occurred during solar system and the Earth Moon system formation. Correlation between the time of impact events based on stratigraphy and the absolute ages of rocks associated with these events has allowed establishment of an absolute time scale for geological evolution, including bombardment activity, in the inner solar system.

The Moon's surface has been primarily shaped and modified by bombardment through time, unmitigated by an atmosphere, resulting in a landscape and a surface layer very different from the Earth's. Extensive use of orbital photography to study the Moon's bombardment-shaped landscape has allowed recognition and resolved controversies about impact features on the Earth.

Lunar soil, the regolith, consists of angular, abrasive fragments, much of it glassy amorphous material known as agglutinates. Fragments vary greatly in size and soil particles average 40–60 μm in diameter. Regolith is distinctively enriched in isotopes implanted by the solar wind, thus recording 4 billion years of solar history useful for understanding variations in Earth's climate through time. Lunar soil lacks organics, and any indication of biological activity.

Consistent with its tidal interactions with the Earth, the Moon is asymmetrical, with a thinner nearside crust. More volcanic activity resulting from impact events on the nearside has resulted in greater mass, resulting from the filling of impact basins with denser more mafic material, and mass anomalies, or mascons, associated with these basins.

References

1. Beatty, D. Taking science to the moon: Lunar Experiments and the Apollo Program. Johns Hopkins University Press. 352 p. 2001.
2. Lindsay, H. ALSEP Apollo Lunar Surface Experiments Package. NASA. 2008. https://www.nasa.gov/history/alsj/HamishALSEP.html. visited 10/1/24.
3. Lewis, H. The Scientific Voice of the Moon, Lunar Surface Exploration. Bendix Technical Journal, 4, 2, 1-10, 1971.
4. McNaughton, J. Some Aspects of ALSEP Structural/Thermal Design. Lunar Surface Exploration. Bendix Technical Journal, 4, 2, 11-19, 1971.
5. Tosh, W. The ALSEP Central Station Data Subsystem. Lunar Surface Exploration. Bendix Technical Journal, 4, 2, 20-27, 1971.

6. Harris, R., Apollo Experience Report - Thermal Design of Apollo Lunar Surface Experiments Package, NASA TN D-6738, 1972. https://ntrs.nasa.gov/api/citations/19720013192/downloads/19720013192.pdf. visited 10/1/24.
7. Bugby, D. and J. Rivera, Thermal Technology advancements for extended-duration lunar operation. 52nd international conference on environmental systems. ICES-2023-244. 2023.
8. Apollo 11 Preliminary Science Report NASA SP-214. 1970. https://ntrs.nasa.gov/api/citations/19700025955/downloads/19700025955.pdf. visited 10/1/24.
9. Apollo 12 Preliminary Science Report NASA SP-235. 1970. https://ntrs.nasa.gov/api/citations/19700025955/downloads/19700025955.pdf. visited 10/1/24.
10. Apollo 14 Preliminary Science Report NASA SP-272. 1971. https://www.nasa.gov/wp-content/uploads/static/history/alsj/a14/as14psr.pdf. visited 10/1/24.
11. Apollo 15 Preliminary Science Report NASA SP-289. 1972. https://www.nasa.gov/wp-content/uploads/static/history/alsj/a15/as15psr.pdf. visited 10/1/24
12. Apollo 16 Preliminary Science Report NASA SP-315. 1972. https://www.nasa.gov/wp-content/uploads/static/history/alsj/a16/as16psr.pdf. visited 10/1/24
13. Apollo 17 Preliminary Science Report NASA SP-330. 1973. https://ntrs.nasa.gov/api/citations/19740010315/downloads/19740010315.pdf. visited 10/1/24
14. Alley, C., et al. Laser Ranging Retroreflector. Apollo 11 Preliminary Science Report NASA SP-214. p. 163–182. 1969.
15. Faller, J. et al. Laser Ranging Retroreflector. Apollo 12 Preliminary Science Report NASA SP-235. p. 215–220. 1970.
16. Faller, J. et al. Laser Ranging Retroreflector. Apollo 14 Preliminary Science Report NASA SP-272. 14-1. 1971.
17. Latham, G. et al. Passive Seismic Experiment. Apollo 11 Preliminary Science Report NASA SP-214. p. 143–162. 1969.
18. Latham, G. et al. Passive Seismic Experiment. Apollo 12 Preliminary Science Report NASA SP-235. p. 39–53. 1970.
19. Latham, G. et al. Passive Seismic Experiment. Apollo 14 Preliminary Science Report NASA SP-272. p. 133–162. 1971.
20. Kovach, R. et al. Active Seismic Experiment. Apollo 14 Preliminary Science Report NASA SP-272. p. 163–174. 1971.
21. Latham, G. et al. Passive Seismic Experiment. Apollo 15 Preliminary Science Report NASA SP-289. 8-1. 1972.
22. Latham, G. et al. Passive Seismic Experiment. Apollo 16 Preliminary Science Report NASA SP-315. 9-1. 1972.
23. Kovach, R. et al. Active Seismic Experiment. Apollo 16 Preliminary Science Report NASA SP-315. 10-1. 1972.
24. Latham, G. et al. Passive Seismic Experiment. Apollo 17 Preliminary Science Report NASA SP-330. 11-1. 1973.
25. Kovach, R. et al. Lunar Seismic Profiling Experiment. Apollo 17 Preliminary Science Report NASA SP-330. 10-1. 1973.

26. Langseth, M. et al. Heat-Flow Experiment. Apollo 15 Preliminary Science Report NASA SP-289. 11-1. 1972.
27. Langseth, M. et al. Heat Flow Experiment. Apollo 17 Preliminary Science Report NASA SP-330. 9-1. 1973
28. Giganti, J. et al. Lunar Surface Gravimeter Experiment. Apollo 17 Preliminary Science Report NASA SP-330. 12-1. 1973.
29. Talwani, M. et al. Traverse Gravimeter Experiment. Apollo 17 Preliminary Science Report NASA SP-330. 13-1. 1973.
30. Dyal, P. et al. Lunar Portable Surface Magnetometer. Apollo 14 Preliminary Science Report NASA SP-272. p. 227–238. 1971.
31. Dyal, P. et al. Lunar Surface Magnetometer Experiment. Apollo 15 Preliminary Science Report NASA SP-289. 9-1. 1972.
32. Dyal, P. et al. Lunar Surface Magnetometer Experiment. Apollo 16 Preliminary Science Report NASA SP-315. 11-1. 1972.
33. Dyal, P. et al. Lunar Portable Magnetometer Experiment. Apollo 16 Preliminary Science Report NASA SP-315. 12-1. 1972.
34. Woolum, D. et al. Neutron Probe Experiment. Apollo 17 Preliminary Science Report NASA SP-330. 18-1. 1973.
35. Bates, J. et al. The Modified Dust Detector in the Early Apollo Scientific Experiment Package. Apollo 11 Preliminary Science Report NASA SP-214. p. 199–292. 1970.
36. O'Brien, B. and M. Hollick, Sunrise-driven movements of dust on the Moon: Apollo 12 Ground-truth measurements. Planetary and Space Science, 119, 194-199, 2015.
37. Berg, O. et al. Lunar Ejecta and Meteorites Experiment. Apollo 17 Preliminary Science Report NASA SP-330. 16-1. 1973.
38. Simmons, G. et al. Surface Electrical Properties Experiment. Apollo 17 Preliminary Science Report NASA SP-330. 15-1. 1973.
39. Fleischer, R. and H. Hart. Cosmic Ray Experiment Part A: Cosmic Ray and Energy Spectra of Solar Cosmic Ray Nucleii. Apollo 16 Preliminary Science Report NASA SP-315. 15-2. 1972.
40. Price, P. et al. Cosmic Ray Experiment Part B: Composition of Interplanetary Particles from 0.1 to 150 MeV/Nucleon. Apollo 16 Preliminary Science Report NASA SP-315. 15-11. 1972.
41. Burnett, D. et al. Cosmic Ray Experiment Part C: Solar Cosmic Ray, Solar Wind, Solar Flare, and Neutron Albedo Measurements. Apollo 16 Preliminary Science Report NASA SP-315. 15-19. 1972.
42. Geiss, J. et al. Solar-Wind Composition Experiment. Apollo 11 Preliminary Science Report NASA SP-214. 183-186. 1970.
43. Geiss, J. et al. Solar-Wind Composition Experiment. Apollo 12 Preliminary Science Report NASA SP-235. p. 99–102. 1970.
44. Geiss, J. et al. Solar-Wind Composition Experiment. Apollo 14 Preliminary Science Report NASA SP-272. p. 221–226. 1971.

45. Geiss, J. et al. Solar-Wind Composition Experiment. Apollo 15 Preliminary Science Report NASA SP-289. 15-1. 1972.
46. Geiss, J. et al. Solar-Wind Composition Experiment. Apollo 16 Preliminary Science Report NASA SP-315. 14-1. 1972.
47. Snyder, C. et al. Solar-Wind Spectrometer Experiment. Apollo 12 Preliminary Science Report NASA SP-235. p. 75–82. 1970.
48. Clay, D. et al. Solar-Wind Spectrometer Experiment. Apollo 15 Preliminary Science Report NASA SP-289. 10-1. 1972.
49. O'Brien, B. and D Reasoner. Charged-Particle Lunar Environment Experiment. Apollo 14 Preliminary Science Report NASA SP-272. p. 193–214. 1971.
50. Hoffman, J. et al. Lunar Atmospheric Composition Experiment. Apollo 17 Preliminary Science Report NASA SP-330. 17-1. 1973.
51. Johnson, F. et al. Cold Cathode Gage (Lunar Atmosphere Detector). Apollo 12 Preliminary Science Report NASA SP-235. p. 93–98. 1970.
52. Freeman, J. et al. Suprathermal Ion Detector Experiment (Lunar Ionosphere Detector). Apollo 12 Preliminary Science Report NASA SP-235. p. 83–92. 1970.
53. Johnson, F. et al. Cold Cathode Gage (Lunar Atmosphere Detector). Apollo 14 Preliminary Science Report NASA SP-272. p. 185–192. 1971.
54. Hills, H. and J. Freeman, Suprathermal Ion Detector Experiment (Lunar Ionosphere Detector). Apollo 14 Preliminary Science Report NASA SP-272. p. 175–184. 1971.
55. Johnson, F. et al. Cold Cathode Gage (Lunar Atmosphere Detector). Apollo 15 Preliminary Science Report NASA SP-289. 13-1. 1972.
56. Hills, H. et al. Suprathermal Ion Detector Experiment (Lunar Ionosphere Detector). Apollo 15 Preliminary Science Report NASA SP-289. 12-1. 1972.

8

Apollo Surface Missions

8.1 Context

Truly, the Apollo program achieved the national goal established by President John Kennedy to 'land a man on the moon by the end of this decade'. As discussed earlier, the Apollo program was the culmination of efforts, involving hundreds of thousands of Americans and our allies around the world, that resulted in arguably the greatest technological achievement of all time: the development of all of the major technologies essential human for deep space exploration in less than a decade. These technologies were and have been applied to many other programs involving robotic missions during that decade and in the ensuing years. In this chapter the emphasis is on specifically how and where the surface exploration methodologies and tools were applied successfully to characterize Apollo landing sites, with progressively greater skill and more sophisticated science goals [1]. The planned but cancelled missions following Apollo 17 are also considered, as are some additional missions baselining Apollo or comparable level mobility technology.

8.2 Early Missions: Apollo 11 and 12

Astronauts Armstrong and Aldrin were the Apollo 11 Lunar Module crew. What were the accomplishments of the very first Apollo mission to the Moon [1–4]? First of all, Apollo 11 provided the first color TV pictures of the Earth during its cruise to the moon, and, of course, verified the capability for an

American-sponsored human landing on the Moon. The Lunar Module also had the first onboard flight computer [5], its guidance and navigation system, and the first flight software, developed by MIT Draper Labs. By today's standards, the computer had limited memory (36,000 words) and speed. Due these limitations, the coding required real cleverness, in some ways more advanced than today's approach to coding, because it had to be extremely compact. Although this system worked well on all of the following missions, the first time around a computer error occurred, not well understood at the time. Later analysis indicated that a switch had accidentally been flipped, turning on the radar, which was using up all available storage space and causing constant restarts and flushing of existing data. Although continually restarting, the computer was still performing critical tasks, so the landing was not aborted. Implementing manual override, Neil Armstrong, recognized as one of the NASA's best pilots, did a manual landing on the Moon. Avoiding a large crater, Armstrong landed 4 miles downrange from the planned target (in Mare Tranquillitatis, a nearside eastern maria, at 0.67°N latitude, 23.47°E longitude), requiring 40 s longer than originally planned and with 17 s of fuel remaining. The selection of the Mare Tranquillitatis landing site was driven by astronaut safety concerns: it was equatorial (easier communication) and smooth (easier landing). Secondarily, it represented one of the two major lunar terranes, volcanic plains (maria).

When the Apollo 11 astronauts arrived on the surface, after that exciting ride, they told mission control that they wanted to advance the schedule: instead of a rest period, they wanted to get out on the surface immediately. The first human being to set foot on the lunar surface, Armstrong made the famous, and controversial statement: "One small step for [a] man, one giant leap for mankind." Did he use the 'a' or not? I believe he did, as the statement makes more sense that way. Armstrong set up the TV camera and took the first sample; Aldrin deployed the Early Apollo Scientific Experiment Package (EASEP) (see Chap. 7): the lunar laser ranging reflector, lunar dust detector, seismic experiment packages, and solar wind composition experiments. Both astronauts collected samples and pictures. As indicated on the Apollo 11 astronaut traverse map (Fig. 8.1), their travels on foot took them up to 50 m away from the lander. Letters on the map indicate sampling stations, stars indicate where photos were taken, and locations of experiments are marked with their acronyms. In the two and a half hours of EVA time they had on the surface, the Apollo astronauts collected 21.6 kg of samples near the lander and utilized 70 mm cameras to take exposures of the landing site. Samples collected during the mission included breccias, impact event-derived rocks consisting of angular, broken fragments cemented by partially melted matrix,

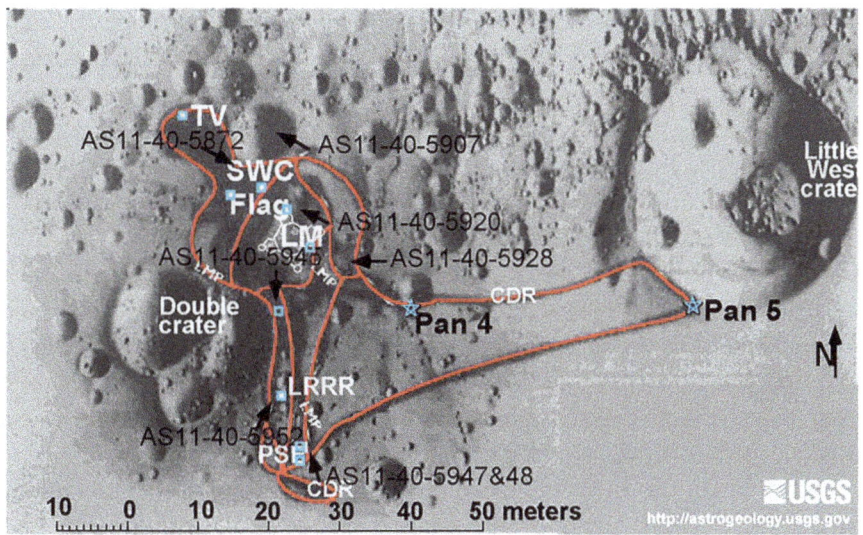

Fig. 8.1 Apollo 11 Landing Site Map. LM and Flag indicate the location of the lander and the flag respectively. The five panoramic camera views (Pan), Hasselblad camera images (Apollo Atlas Image numbers beginning with AS11), the TV camera (TV), and EASEP instrument (SWC, LRRR, PSE, CDR) deployment sites indicated. Two major landmarks, Double and Little West Crater, are also indicated. (Courtesy of USGS Astrogeology Branch)

basalts derived from volcanic flows of Mare Tranquillitatis, and fragments of surrounding highland rocks. The basalts provided the basis for the idea that the earliest lunar surface was entirely molten, a 'magma ocean', where less dense highland rock rose to the top (to form the highlands) and more dense minerals that formed the source region for mare basalt sunk.

Astronauts Conrad and Bean were the Apollo 12 Lunar Module crew. For the Apollo 12 mission, the flight computer was upgraded with real time correction capability to allow more precise landing. The Apollo 12 landing site, in Oceanus Procellarum (at 3.01° South latitude, and 23.42° West longitude) was not only selected because it was 'safe': equatorial, smooth, and level. The site could also provide access to Surveyor III, the third unmanned soft lander, allowing the astronauts to retrieve portions of the lander to see how well it had survived and to see the inside of the crater, the most ubiquitous landform on the Moon, where Surveyor had landed. A precision landing was achieved: the lander was 500 feet from Surveyor!

What were the primary activities of the Apollo 12 mission [1, 6, 7]? Conrad collected the first lunar samples and deployed the S-band antenna. Experiments of the first ALSEP package, which included a central power package, were deployed, including the solar wind experiment, passive seismometer, lunar surface magnetometer (LSM), lunar dust detector, and SIDE/CCIG experiments. For the first time, moonquake activity was detected with the seismometer, providing a great advance in our understanding of the lunar interior. Bean attempted to deploy the TV camera, but it was inadvertently pointed toward the sun and 'fried'. The issues with 'stiffness' of the power cables connecting experiments to the central power station were rectified for future ALSEP packages. As indicated on the Apollo 12 traverse map (Fig. 8.2) the astronauts visited Surveyor III to retrieve portions of the lander, walking up to 500 m away from the lunar module. During their seven and a half hours of EVA time on the surface, the astronauts collected 34.5 kg of samples, most of them near the lander, and the first core sample, and utilized 70 mm cameras to take exposures of the landing site.

Largely basaltic samples indicated the mafic volcanic origin of the underlying Oceanus Procellarum and the relatively young age of the underlying mare formation. The small number of light fragments found indicated the presence of another rock suite of noritic (intermediate) composition, considered to be

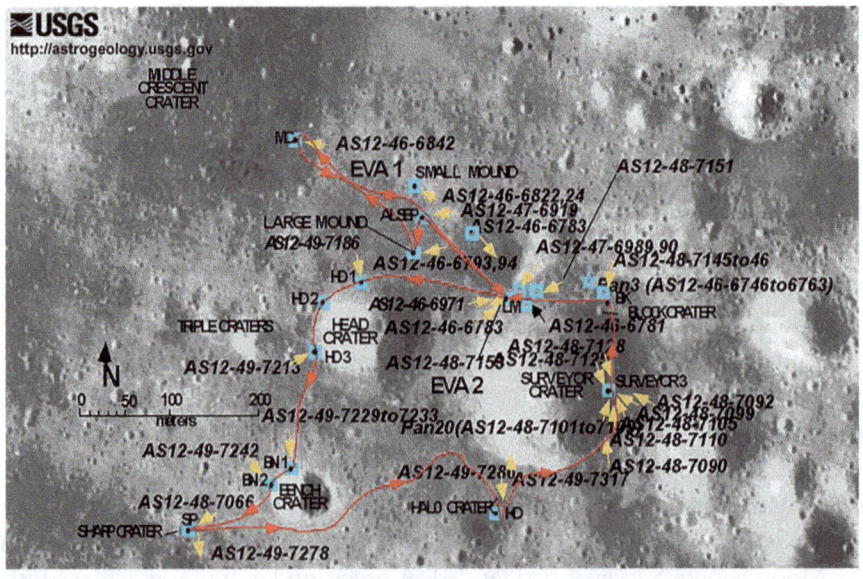

Fig. 8.2 Apollo 12 Landing Site Map. Locations of the lander (LM), ALSEP, and Surveyor 3 are indicated, along with major physical landmarks labelled. The EVA1 and EVA2 traverses follow the red lines. Panoramic camera views (Pan), Hasselblad camera images (Apollo Atlas Image numbers beginning with AS12), HD (high definition), MD (medium definition), photos, and local photos (SP, HD, BK, BN) for Sharp Crater, Halo Crater, Block Crater, and Bench Crater) indicated. (Courtesy of USGS Astrogeology Branch)

a residual liquid remaining after the formation of the lunar anorthositic (highly aluminous) crust.

8.3 Transitional Mission Apollo 14

Apollo 14 Astronauts Shepard and Mitchell landed on the Fra Mauro Formation, at 3.64° South latitude, and 17.47° West Longitude, midway between Doublet and Triplet Craters in the hilly uplands of Fra Mauro. Despite problems with a short circuit in the abort switch and landing radar, they landed within 30 m of the target. Their traverses are indicated on the accompanying map (Fig. 8.3). What were their accomplishments [1, 8–10]? During the first EVA they deployed the ALSEP, then collected rocks and soils. A communication problem cut short the first EVA. During the second EVA, they walked to Cone Crater, using the Modular Equipment Transport to carry

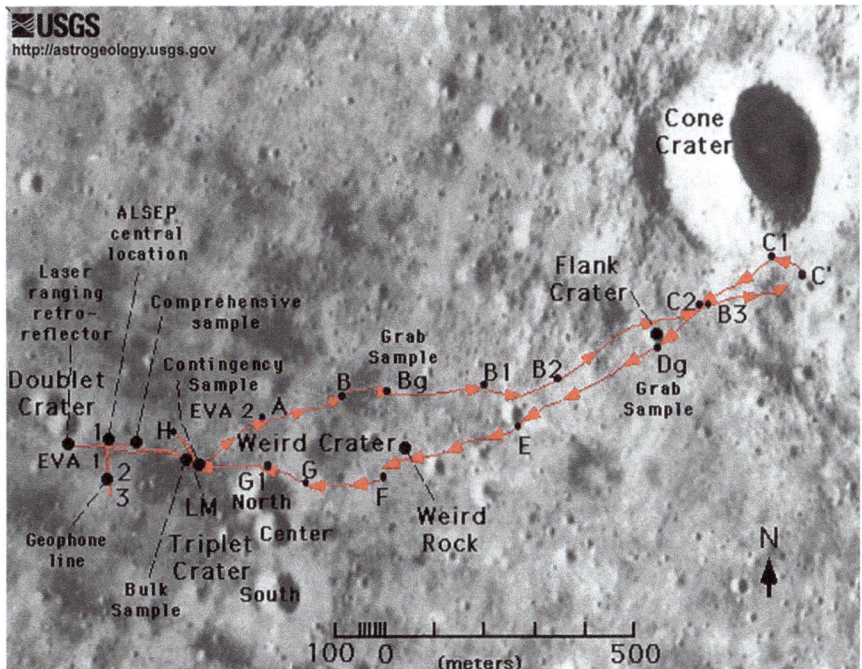

Fig. 8.3 Apollo 14 Landing Site Map. Locations of the lander (LM), ALSEP with Geophone line, LRRR, Bulk, Comprehensive, Grab, and Contingency samples are indicated, along with major physical landmarks labelled. The EVA1 and EVA2 traverses follow the red lines. A, B, C, D, E, F, G, and H indicate documentation taken sequentially along the second traverse. (Courtesy of USGS Astrogeology Branch)

equipment and collected samples. Although wheeled, the Transport did not work well in lunar regolith, and they ended up carrying it most of the time. Near Cone Crater, the very hummocky terrain and large boulders slowed their progress. They spent about nine and a half hours on the lunar surface during their two EVAs, using a 70 mm camera, 60 mm Hasselblad cameras, and 16 mm Maurer data acquisition cameras to collect black and white and color exposures of the lunar surface. Their sample collection and documentation of the landscape of the Fra Mauro Formation on foot provided the evidence that it consists of rocks that were ejected and emplaced during the impact event that formed the Imbrium Basin, a major event in lunar history. 1100 km Imbrium is the second largest confirmed basin and one of the youngest on the Moon. From Apollo 14 rock samples, scientists were able to determine that the impact event occurred 3.8 billion years ago. The astronauts also found impact breccia and melts associated with older (>4 billion years old) basalts.

8.4 The Apollo J Missions

The introduction of the unpressurized rover and extended life support equipment on Apollo 15, 16, and 17 (the so–called J missions) allowed the most advanced versions of the Apollo methodology to be implemented on those missions [11–15]. In addition to site characterization and sampling activities, all missions included objectives to emplace and activate an automated science experiment package that would characterize local, regional, and subsurface structures and environments as well as to continue reconnaissance for future missions, and to further investigate the lunar surface and near moon environment from orbit. Each mission included a brief preliminary assessment from the lander just after arrival, and three 6–8 h EVAs in each of the 3 days on the surface. The first half of EVA 1 typically involved deploying most of the science experiment package. The bulk of EVA time was used for three traverses designed to characterize and sample the targets of major geological interest (last half of traverse 1, traverse 2, typically the longest, and traverse 3 on the day of departure).

Pre–landing, initial post–landing, and later matured interpretation for Apollo 15 [16–25], 16 [20–22, 26–36] and 17 [20, 21, 37–45] are discussed here to illustrate the successes, and maturing process of the Apollo field methodology .

8.5 Apollo 15 Mission (Figs. 8.4 and 8.5)

The area of interest (Apennine Mountains Front/Mare Imbrium Edge/Hadley Rille) is located (24.5°N, 2.5°E) at the interface of two major terranes: a mare flood basalt filled embayment at the edge of the Mare Imbrium basin called Palus Putredinis and the highland crustal massifs (Apennine Ridge) surrounding it. This was apparently an area of intense mare activity at the edge of Hadley Rille. This landing site was chosen for the access to a wide range of materials from the two terranes as expressed in the great variation in photogeological units. What was not anticipated was the finding of not just more diverse mare basalts, but two new volcanic rock types, including the pre–mare KREEP (high potassium (K), Rare Earth Elements (REE), and phosphorus (P)), and green glass, an ultramafic pyroclastic [23]. Astronauts Scott and Irwin collected 370 regolith and rock samples (77 kg) and a 2.4 m deep drill core.

The initial field science objectives of Apollo 15, successfully accomplished [16, 17], were to investigate, survey, photograph, and sample a sinuous rille, and its major materials and structures, as well as Apennine materials in the landing area that were potentially impact ejecta from the Imbrium Basin (Hadley Delta). The Apennine Front (Hadley Delta Scarp) was visited on both the first and second traverses. The mountains were rounded with gentle to moderate slopes and sparse cratering and debris, thought to have been gravitationally transported to the lower slopes. Unanticipated were the sharp, deeply etched, parallel lineations on the mountain faces, representing either underlying bedding or regional fractures. A dark band near the base at nearby Mt. Hadley was interpreted to represent the remnant of a high–lava mark left after subsidence during an episode of basin–filling volcanism. Hadley Rille, thought to be one of the youngest sinuous rilles on the Moon, was visited on both the first and third traverses. Its underlying structure, with strata of greatly varying thickness, albedo, and textures representing individual lava flows exposed on the upper walls, and blocks far larger than those observed on Earth broken away from the walls on its lower talus slopes, were extensively documented and sampled at several large outcrops. All the traverses crossed the smooth to gently undulating mare surface punctuated with younger craters and their rougher ejecta blankets. Four distinct geological units could be discerned in the mare area based on differences in crater population and surface texture.

Experiments deployed in the Apollo 15 ALSEP (Apollo Lunar Science Experiment Package) included (1) the passive seismic experiment (deployed

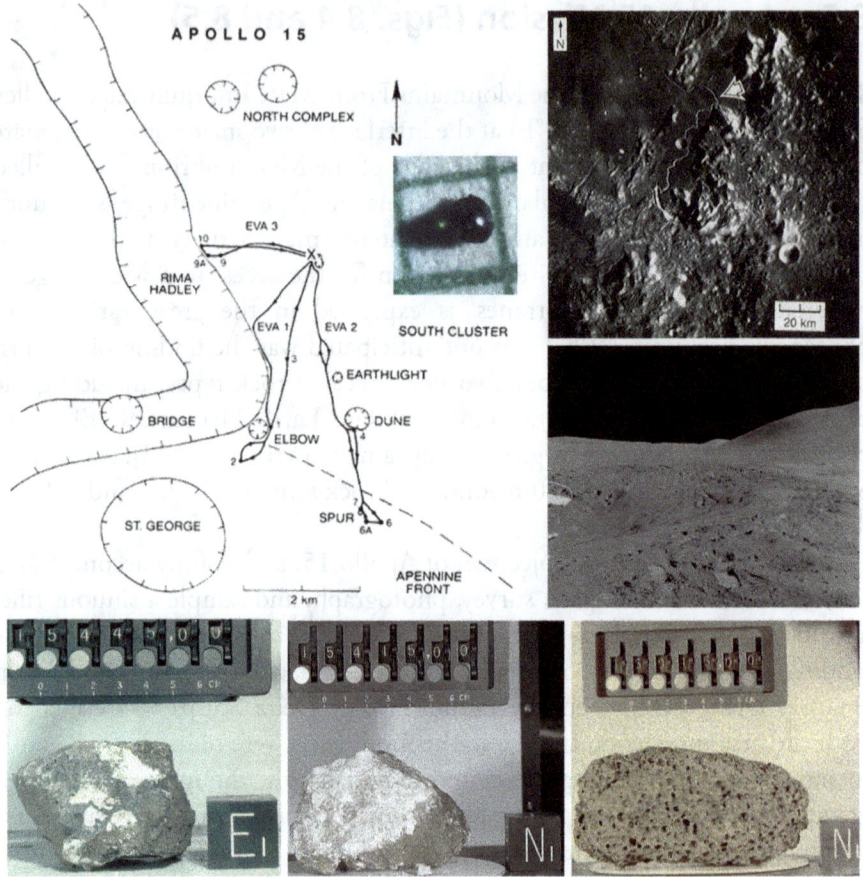

Fig. 8.4 Apollo 15 traverses (left), orbital image (top right) and ground photo (middle right) of landing site. Most significant samples include (left to right along bottom) Imbrium basin impact melt rock, the oldest piece of crust collected called 'genesis rock', and young very low titanium flood basalt, as well as (insert above), the first pyroclastic material discovered, green glass [11]. (Courtesy Clark, Fig. 62,010, Earth Moon and Planets)

on previous mission), (2) the stationary surface fluxgate magnetometer (deployed on previous missions), (3) the heat flow probe, (4) the laser ranging retroreflector (for determining Earth–Moon distance and system dynamics) (deployed on previous missions), (5) the Solar Wind Spectrometer (for determining the densities and velocities of solar and magnetospheric plasma) (also deployed on Apollo 12), (6) the Solar Wind Composition Experiment (deployed on previous missions), (7) the Suprathermal Ion (lunar ionosphere) Detector (deployed on previous missions), (8) the cold cathode gage (lunar

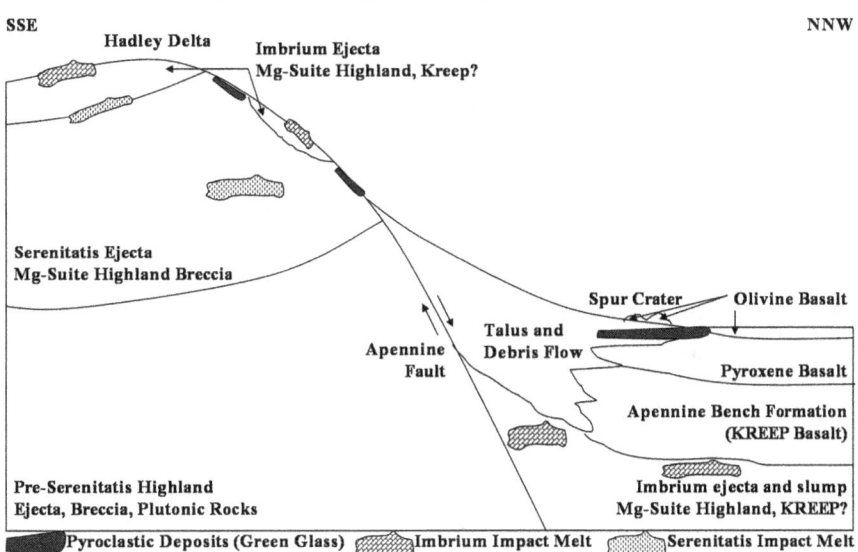

Fig. 8.5 Schematic Geological Cross–Section of the Apollo 15 Landing Site Area from SSE to NNW (modified from [22]) illustrating the complex nature of the interface between underlying crustal structures generated by impact represented by Hadley Delta, crustal impact debris generated by several major impacts, and episodic volcanic plains formation subsequent to Imbrium basin formation [11]. (Courtesy of Clark, 2010, Earth Moon and Planets)

atmosphere detector) (deployed on previous missions), and (9) the soil mechanics experiment (deployed on previous missions).

The landing site [16–20, 23] is a valley at the foot of the Apennine Front, uplifted crustal fault blocks of underlying highland crust which rise 2–5 km above the floor of Palus Putredinis to form the rim of the Imbrium basin. For scale, this fault scarp rises higher than the Himalayan Front above the plains of India. The moderate to steep slope front at Mount Hadley grades into low hummocky hills to the east and a line of hills across a mare band to the west. The many passes formed between the hills are in some cases filled with mare flood basalt, and in other cases with debris traveling downslope. The Apennine Mountains, thought likely to be ancient material exposed during the excavation of the Imbrium Basin, were of great interest as potentially representative of crustal material underlying the other mare basins. The site is just east of Rima Hadley, a 1-km wide, 400-m deep v-shaped sinuous rille apparently originating at a volcanic elongated depression in an area of domes in the Apennines and essentially paralleling the Mountain front along the eastern boundary of Mare Imbrium. The mare surface sloped and the regolith thinned toward the rille, providing greater opportunity to sample underlying bedrock. Questions about the role of volatiles, the origin of the fluid and nature of the flow mechanism involved in its apparently volcanic formation, made the rille an object of great interest; however, its 25–30° wall slopes, discontinuous outcrops of stratified beds just below the rim, talus slopes to the bottom, and blocky debris on the floor, made travel into the rille too risky. Nevertheless, samples collected confirmed its volcanic origin as a lava tube. On the east, the valley is cut off by a gently sloping 80 m scarp that becomes a dome–shaped wrinkle ridge farther north.

What was the nature of materials collected here and how diverse and representative of underlying materials were the samples [21, 22, 24, 25]? Rocks typically collected at the front were breccias (impact fragments more or less welded by impact melt): friable with inclusions of mare basalt, KREEP basalt, and/or glass; coherent with a glassy matrix with inclusions of KREEP basalt and ferromagnesian mineral (olivine and pyroxene) grains; or more lithified with aluminosilicate (feldspathic) mineral (plagioclase feldspar) inclusions. Rocks from the rille and surrounding mare were typically olivine and pyroxene rich basalts. The oldest known lunar rock, and one of the youngest rocks found, a very low titanium basalt, are represented in the Apollo 15 rock collection, along with the first example of volcanic (pyroclastic) green glass. One of the oldest chunks of highland crust found on the Moon came from the Apennine Ridge. The most mare basalt rich soil was found at the edge of Hadley Rille. Mare soils were a mixture of olivine (olivine bearing) and quartz (pyroxene bearing) normative basalts of the same ages, older underlying KREEP basalt, and younger ultramafic pyroclastics. Mare soil composition could be made to match underlying bedrock (as indicated by rock samples) with the addition of magnesium rich mafic plutonic rock (gabbro) and KREEP basalt. Samples from the Apennine massifs consist of high–magnesium felsic highland soils and breccias generated as impact debris during pre–Imbrium (Serenitatis) basin formation and then transformed into basin rim massifs during the Imbrium impact event. The correspondence between rock (representing underlying bedrock) and regolith samples is less pronounced at Apollo 15 than at other sites most likely due to the combination of less pronounced compositional variations and greater mixing and transport of materials [22].

8.6 Apollo 16 Mission (Figs. 8.6 and 8.7)

The area of interest [26–28] was the Descartes Plains lying in the southeastern highlands of the nearside, southwest of Mare Tranquillitatis. The landing site was located (9.0°S, 15.5°E) just north of Descartes Crater in the Descartes

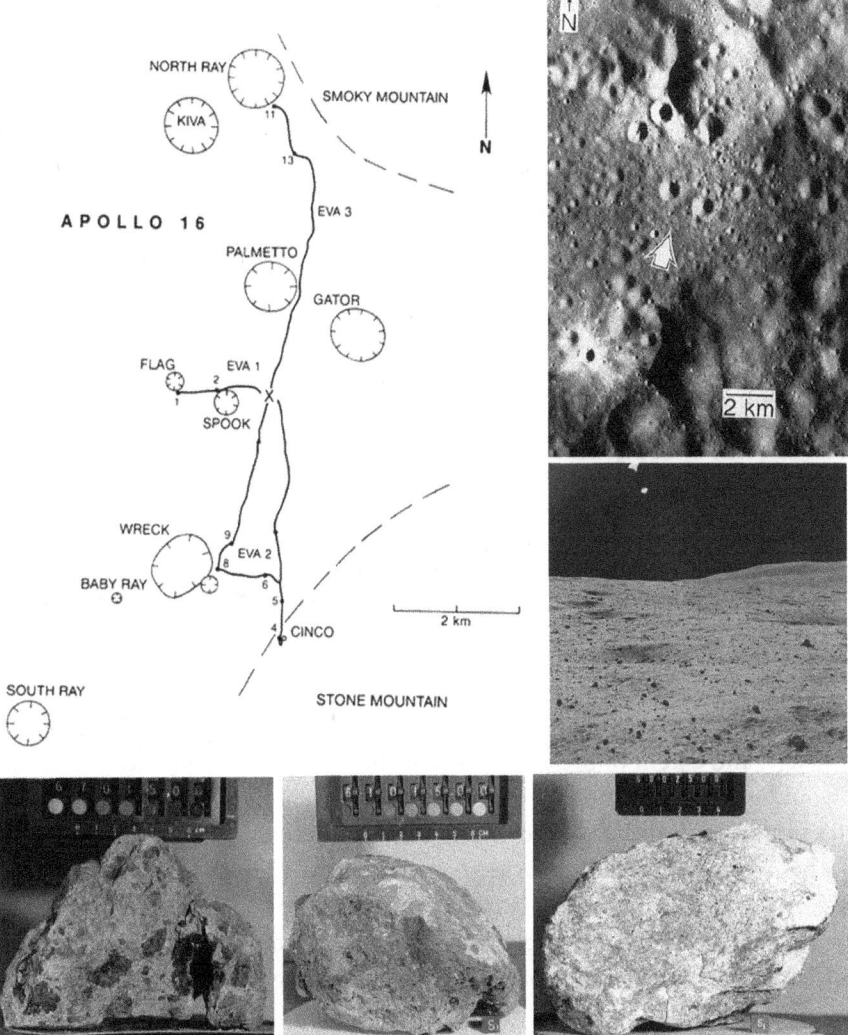

Fig. 8.6 Apollo 16 traverses (left), orbital image (top right) and ground photo (middle right) of landing site. Typical rocks, include breccia (left) with clasts containing dark volcanic material, impact melt (middle), and bright crustal fragments (extreme right) [11]. (Courtesy of Clark, Fig. 8, 2010, earth moon and planets)

formation, a hilly, grooved, furrowed terrain associated with ridges and ramparts and apparently morphologically similar to many terrestrial areas of volcanism. The site was also adjacent to the Cayley formation, highland plains material thought to be upland basin fill consisting of higher albedo welded tuff and analogous to volcanic mare basin fill. Although initially interpreted to be of impact origin, these two formations were subsequently thought, by overwhelming consensus, to be highland volcanic complexes analogous to

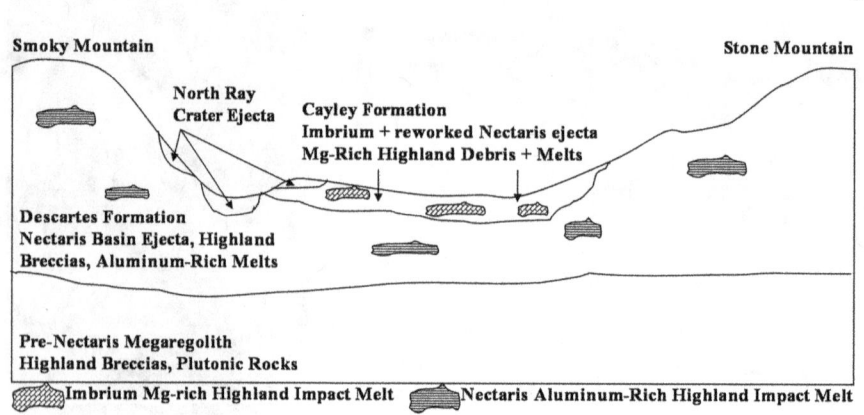

Fig. 8.7 Schematic Geological Cross–Section of the Apollo 16 Landing Site Area from N to S (modified from [22]) illustrating the complexity introduced by repeated excavation and deposit of impact ejecta and melt of highland crust, as well as the stratigraphic relationship between the younger Cayley Formation, the surrounding Descartes Formation, and an even older underlying formation [11]. (Courtesy of Clark, Fig. 9, 2010, earth moon and Planets)

mare volcanic complexes which would provide constraints on the composition, age, and extent of pre-mare volcanism. The site was also selected because it would provide access to the most 'pristine' highlands furthest removed from sites of mare volcanism and to young impact structures.

The Apollo 16 mission provided a real surprise [22, 28–33, 35] to the science investigation planners, because, as a result of their field training, the astronauts recognized relatively quickly that the volcanic interpretation was not correct. They found impact breccias indicating that the area was not of volcanic origin but was dominated by impact debris. The orbital X-ray data supported an impact origin for the Apollo 16 as well, indicating that the landing site had a composition representative of large areas of the nearside highlands. As a result of the 'surprise' provided by Apollo 16, photogeological interpretation for future mission planning was augmented by the inclusion of multiple interpretations based on multiple working hypotheses, and many highland photogeological units previously considered volcanic were afterward interpreted to be of impact origin.

The initial field science objective involved investigating, surveying, and sampling materials and features representing the Descartes (and Cayley) formations [27, 28]. The Cayley formation was sampled extensively from north to south during all the traverses. Based on panoramic photography of South Ray Crater, the formation consists of rough, horizontal layering of light and dark breccia with impact textures apparently derived from (crustal) plutonic anorthosites and anorthositic gabbros. At the southern end of the second traverse, the Descartes formation photogeological unit was sampled at Stone Mountain; yielding rocks and soils with compositions and textures indistinguishable from Cayley material, indicating that the upper layers of the Descartes formation may be from the same source (crustal highlands) as the Cayley formation. All traverses were also used to explore local impact ejecta from craters of different sizes, thus sampling underlying stratigraphy to different depths. Apparently, the area is covered with breccia from the older Descartes (Nectaris Basin ejecta) and the younger Cayley (Imbrium Basin Ejecta) Formations. Astronauts Young and Duke collected 731 rock and regolith samples (95.7 kg) and a 2.2 m deep drill core.

Experiments deployed in the Apollo 16 ALSEP included (1) and (2) passive and active seismic experiments (deployed on previous missions), (3) and (4) stationary surface fluxgate and portable magnetometers (deployed on previous missions), (5) the Far UV Camera/Spectrograph (Earth and astronomical object observatory), (6) the Solar Wind Composition Experiment (deployed on previous missions), (7) the cosmic ray detector, and (8) the soil mechanics experiment (deployed on previous missions).

The Apollo 16 landing site [30–33, 35, 36] consists of younger Imbrium impact debris plains made from highland crustal material, the Cayley Formation, adjacent to the older Descartes formation to the south. The plains form a valley between older surrounding hilly and furrowed mountainous terrains, Smokey Mountain, a steep sided plateau to the north, and Stone Mountain, a gently sloping boulder covered hill to the south. North and South Ray craters are young and thus well preserved, their walls with up to 45° slopes consisting of outcrops, boulders and thin regolith. South Ray, the youngest of the two, has blocky ejecta near the rim. All terrains are heavily cratered.

Apollo 16 soils [21, 36] closely approximates the typical highland rock, anorthositic gabbro thought to be most representative of the pre-mare lunar crust [34]. Soils match local bedrock composition, as indicated by collected rock samples, when an anorthositic (most felsic highland rock) component is added. Collected samples show the least variation compared to other Apollo missions, at least in part due to the homogenizing of bulk composition that occurs during the formation of crystalline breccias.

8.7 Apollo 17 Mission (Figs. 8.8 and 8.9)

The area of interest [37–39] was a complex old basin edge highland/mare/young impact feature interface analogous to the Apollo 15 site, the Taurus Littrow Valley. The landing site was located (20.16°N, 30.75°E) in low-lying dark volcanic plains south of the crater Littrow nestled between two massifs of the southwestern Taurus Mountains that form the rim of southeastern Mare Serenitatis. The Taurus Mountains were thought to consist of blocks of pre–Imbrium crust faulted and uplifted during Serenitatis Basin formation. Massif materials, potentially exposing deep underlying crust, and younger ejecta could be sampled easily from a major debris flow originating from a

Fig. 8.8 Apollo 17 traverses (left), oblique orbital image (top right) and ground photo (middle right) of landing site. Typical rocks include (bottom left to right) crustal (highland rocks) like olivine–bearing troctolite (left), norite (middle), and the pyroclastic find (orange glass) [11]. (Courtesy of Clark, Fig. 10, 2010, earth moon and planets)

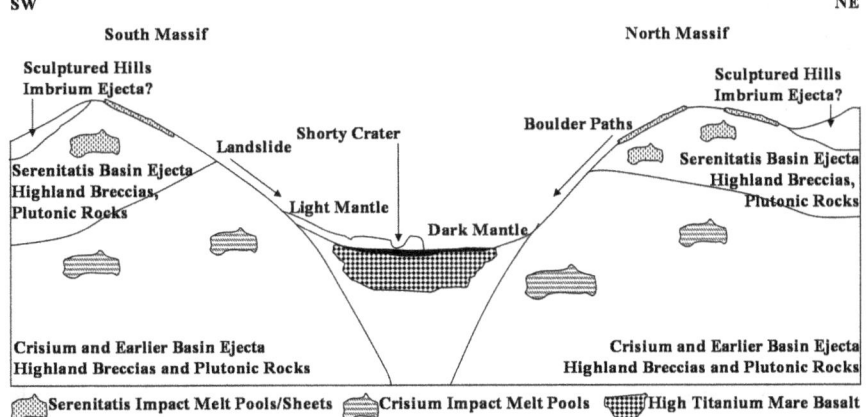

Fig. 8.9 Schematic Geological Cross–Section of the Apollo 17 Landing Site Area from SW to NE (modified from [22]) illustrating the complexity of the interface between mare filled graben Littrow Valley and surrounding mountains of Serenitatis rim, as well as the distribution of the diverse materials found in the area, including light mantle (South Massif landslide) and dark mantle overlying older mare basalt in the valley, crustal ejecta and breccias of the massifs, and the Sculptured Hills material [11]. (Courtesy of Clark, Fig. 11, 2010, earth moon and planets)

massif just southwest of the landing site the dark to very dark lowland valley between the massifs was interpreted as very young volcanic pyroclastic mantle originating deep within the Moon and covering partially exposed earlier mare formations beneath. The area is complex, and potentially challenging to interpret, with surface units and rocks varying greatly in appearance, apparent age, composition, and origin. This mission was the best planned and executed because it could build on so much previous experience and take advantage of the presence of a geologist on the Moon [22, 39–44]. The preliminary interpretation of Taurus Littrow as a graben valley formed at the time of the Serenitatis Basin, later reactivated with episodes of basaltic volcanism, were largely, though not entirely, supported by the field investigation. The completely unanticipated find was the orange glass, a pyroclastic material initially misinterpreted as indicating the presence of oxidation products.

The primary field objectives were to investigate, survey, and sample the wide range of surface units and materials of the Taurus Littrow region, including light mantle, dark mantle, and subfloor units, massifs, and Sculptured Hills [38, 39, 44]. Light mantle material was sampled during Traverse 2 at stations associated with the south massif. It consisted of fine-grained debris with breccias similar to those associated directly with the South Massif. The material was thought to have originated from a secondary impact-induced slide on the north face of South Massif. Lighter mantle material also included a young ray deposit from Tycho Crater. The so-called dark mantle material covering the valley floor, sampled during all the traverses, turned out to be not younger ejecta from impact into surrounding volcanic material but regolith derived from the underlying basalt unit, locally mixed with pyroclastics from cinder cones. The subfloor unit, sampled from boulder-size ejecta from large craters nearest the lander, consisted of an older light-colored (greater proportion of anorthositic mineral assemblages) basalt thought to underlie the surface units. Volcanic materials were typically higher titanium basalts. The low slopes of the Sculptured Hills were sampled during Traverse 3 and found to be a highland unit distinctively different from the North and South Massifs, which were sampled during Traverses 3 and 2, respectively. Boulders sampled at the Massifs, which apparently rolled down from higher up the slopes, consisted of a complex range of breccias originating from more than one impact event, as in Apollo 15 and 16.

Experiments used in the field or deployed in the Apollo 17 ALSEP included (1) and (2) passive and active seismic experiment (similar experiment deployed on previous missions), (3) the heat flow probe (deployed on previous missions), (4) and (5) traverse and stationary surface gravimeters, (6) surface electrical properties experiment, (7) the lunar atmosphere composition experiment (deployed on previous missions), (8) the lunar neutron probe (to determine subsurface composition), (9) the Lunar Ejecta and Meteorites experiment, (10) the cosmic ray detector (deployed on Apollo 16), and (11) the soil mechanics experiment (deployed on previous missions).

The Apollo 17 site [20, 40–44] consisted of nearly flat, heavily cratered mare plains surrounded by basin rim forming massifs rising 2–3 km above the valley floor at the margin of Serenitatis. Craters covered the full range of sizes and ages and included secondary clusters. Younger pyroclastic glass form dark mantle material on the older flood basalt deposits. The massifs have moderate to steep slopes unstable enough to lead to mass wasting as evidenced by boulders with long (up to 2 km) tracks descending from outcrops and an avalanche deposit from the South Massif covering a portion of the valley floor and burying older craters beneath with light mantle. The Sculptured Hills form a third terrane distinct and separate from the massifs. The hills are concave, with upper slopes of 20–27° and lower slopes of 15–20°.

How did the nature and diversity of materials collected at Apollo 17 compare to materials found at other Apollo sites [21, 22, 45]? One of the most exciting moments of the mission involved the discovery of pyroclastic orange glass on Shorty Crater during the second traverse. The glass was approximately the age of the younger basalt unit. The largest variety of crustal rocks, including norite, troctolite, and anorthosite, were found at this complex site. Massif soils apparently have a KREEP component. Soil compositions showed the greatest contrast (between mare and highland types) and most closely resembled and were controlled by underlying bedrock composition. The surface visual and compositional variations truly can be used to represent local bulk composition. Underlying crust is best represented by complex breccias and plutonic magnesium rich plutonic rocks, the oldest of which may have originated from the Serenitatis impact event. The underlying mare deposit consist of high titanium basalts with an age of 3.7–3.8 billion years. The pyroclastic orange glass deposits have a younger age of 3.64 billion years. The North Massif is more gabbroic than the South Massif. The Sculptured Hills composition seems to be represented by equal portions of soils from all the distinctive compositional units, including valley floor, dark mantle pyroclastics, and highland massifs.

8.8 Understanding and Integrating Apollo J Mission Results

As the Apollo landings clearly indicate, on the Moon, the crater is the most ubiquitous landform. Natural Excavation overturns and reverses bedding. Basalt flows flood basins to extent which depends on depth and formation age. Impact and Volcanic activity create horizontal stratigraphy discernable with traverses. Radial and Concentric structural patterns are associated with both volcanic and impact events. Thus, the basic geological field approach [46, 47], determining temporal relationships from spatial relationships using the superposition principle, is applied to the Moon by observing such relationships for exposed strata in impact or volcano-tectonic formations, in surface deposits, such as ejecta or pyroclastic deposits, or by traversing regolith for expression of underlying geochemical boundaries, such as flow fronts. Surface Expression of underlying stratigraphy will be found in discernable stratigraphic contacts and unconformities, both vertical (crater, volcano-tectonic ridge, scarp, or rille walls) and horizontal (traverse underlying bedding, sample surface outcrops, cross volcanic flow fronts or impact debris fields). Ground Truth is currently limited to a small number of nearside, mostly near equatorial sites, extended by remote sensing and imaging with relatively low spatial resolution (by terrestrial standards) with obvious implications for considering 'multiple working hypotheses'. The context for developing and testing hypotheses involves combining observations and direct samples from field work at landings sites limited in geographical extent, with major contributions from global remote observations and terrestrial analogue studies.

The samples and measurements collected using the Apollo field methodology not only characterized the landing sites and illustrated, by implication, the great degree to which local geological processes were globally distributed on the Moon, but revolutionized our understanding of planetary formation and evolution [48]. These processes included basin formation with associated concentric and radial fault systems through major impact events; impact deposits, impact melt, and regolith generation through impact events at every scale; and volcanic processes including basin filling flood basalt volcanism, pyroclastic activity, cone and dome formation, and rille formation. Samples returned to Earth were subjected to geochemical analysis and radiochemical dating. These data combined with measurements from the instruments and tools led to the development of a model for lunar, and by implication other rocky terrestrial body, formation that became known as the now familiar magma ocean model. Geochemical differentiation of the interior formed the aluminosilicate rich crust and the mafic (eventual) mare basaltic source regions at the mantle boundary, with residual material (e.g., lunar KREEP), remaining between crust and mantle. An early, relatively short period bombardment by the greater number of larger projectiles left over after planet formation, led to the large basin formation. This was followed by eruption of flood basalts into the basins, with the compositional character and style of volcanism within a basin and between basins evolving as a function of time. The period of the most intense lunar geological activity was over more than 3.5 billion years ago, turning the Moon into a 'time machine', that gave evidence of what the Earth Moon environment was like in the earliest solar system, and then recorded regolith–generating smaller meteoritic bombardment ever after.

8.9 The Cancelled Missions: Apollo 18, 19, and 20

Interest in the Moon did not end with the Apollo 17 mission, as clearly exemplified by the ongoing fascination with the American moon landings on the part of the public all over the world to this day. What happened to Apollo 18, 19, and 20?

It is certainly true that NASA was under pressure due to the greater financial commitment its remarkable accomplishments exacted than originally planned and the escalating expenses of the Viet Nam War at the time. But, the hardware for the remaining Apollo missions, including the enormously expensive Saturn V's, was already built. All along, NASA had had plenty of naysayers and detractors among the political and academic elites who, from day one, could not possibly conceive of the value of human exploration of the Moon to America's economic health or strategic future outside of 'winning the cold war', and were resentful of the investment in human space exploration and its popularity with the American public. It is a testament to strength of the commitment to the Apollo program on the part of those testifying to congress about the tragedy of the Apollo 1 fire that NASA was able carry on with Moon landings at all after that. Meanwhile, after the particularly emotion-laden incident with Apollo 13, and its potential to fulfill the media approach to coverage evolving with front line TV coverage of the Viet Nam War based on the mantra 'if it bleeds it leads', the major media became bored with the risky business of space travel as 'too routine'. These factors combined with President Nixon's simmering resentment of one of Kennedy's highest profile accomplishments led Nixon to terminate what was probably the greatest ongoing technological achievement of all time, seriously undermining a critical element in four decades of aerospace engineering capability in terms of 'community of practice'. Notably, Nixon initiated this action right after speaking to the Apollo 11 astronauts on their successful return. No, the Apollo program was not cancelled so the Saturn V built for Apollo 20 could be used for Space Lab. Space Lab was a way for NASA to extract something useful from an otherwise major loss.

The value, that the elites couldn't see, included both the technological advances themselves, deliberately made readily available for 'transfer' to industry for profit readily for no to low cost (a very different model from the one that exists in other government agencies) as well as the prestige of being first at those achievements. (Note: Economic studies indicate that every dollar spent for NASA's Apollo program resulted in approximately $10 generation of new revenue.) In retrospect, particularly by initiating the trend of microminiaturization seen in consumer electronics, the Apollo program and its essential predecessors laid the foundation for American leadership in creating an aerospace-based sustainable economic infrastructure. Only recently is this vision being realized. The emergence of microminiaturization-based disruptive technologies for low-mass, low-cost spacecraft (epitomized by the CubeSat paradigm) was embraced by NASA in its 2016 Lunar Exploration Initiative to capture the new high ground: cislunar space. The implications or future lunar exploration, with or without humans in the loop, will be discussed further in Chap. 9.

What would Apollo 18, 19, and 20 have been like? What were the priorities for landing sites? The original recommendations for missions were [49, 50]:

- H-2 (Apollo 13) Fra Mauro
- H-3 (Apollo 14) Littrow
- H-4 (Apollo 15) Censorinus
- J-1 (Apollo 16) Descartes
- J-2 (Apollo 17) Marius Hills
- J-3 (Apollo 18) Copernicus
- J-4 (Apollo 19) Hadley
- J-5 (Apollo 20) Tycho

After the Apollo 13 incident, Apollo 14 was rescheduled to Fra Mauro, Apollo 15, 16, and 17 were assigned as discussed in previous sections, and Apollo 18, 19, and 20 were assigned to Aristarchus/Schroter's Valley, Marius Hills, and Copernicus, respectively.

The cancelled mission landing sites were arguably the most challenging and 'grand' [51, 52]. Marius Hills has the highest concentration of volcanic features on the moon including domes, cones, rilles, channels, and flows representing the Moon's major mare volcanism episodes. Tycho is the most prominent young impact crater visible from Earth. Copernicus crater, with its prominent ray system, is typical of the Copernican period. Also under consideration were Hyginus Rille, due to its most prominent linear rille, Censorinus with its high albedo crater, and, proposed by Apollo 17 astronaut and geologist Jack Schmitt, the mysterious Tsiolkovsky crater, the only farside crater filled with mare basalt.

Apollo 18 (Fig. 8.10): Aristarchus Plateau, the most diverse region on the Moon, has its longest and deepest canyon, the collapsed lava tube Schroter's Valley, and highland outcrops at the edge of the valley. The focus would have been on determining the duration and intensity of volcanic eruptions associated with collapsed lava tube. Likely pyroclastic deposits had already been identified on the plateau. Cobra Head, which Apollo 15 astronauts identified as a volcanic vent, already increased the likelihood of finding other volcanic vents. Transient Lunar Phenomena, local short-lived changes in appearance thought by some to indicate outgassing from volcanic vents, had already been observed in the area. Planners had identified a landing site 2 km from

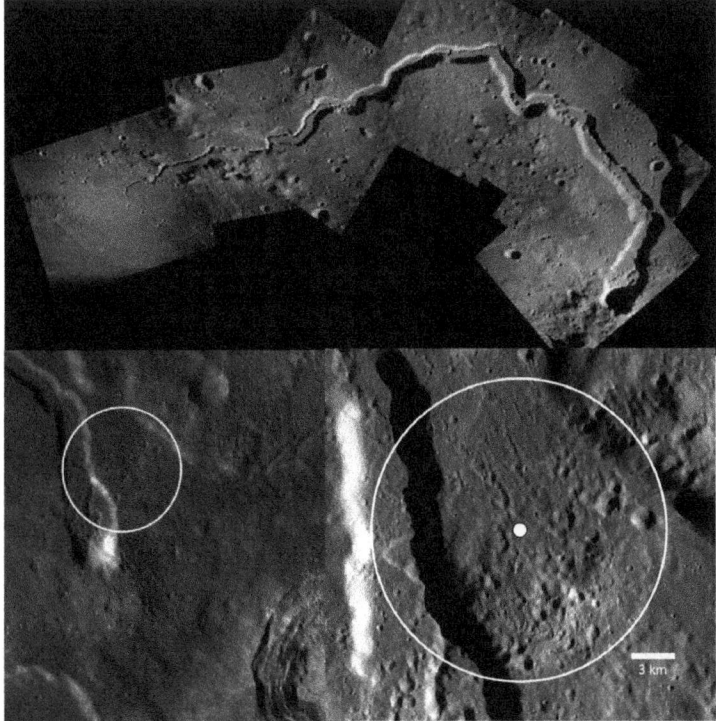

Fig. 8.10 Apollo 18 Proposed Landing Site Photos. Top: Schroter's Valley as described I the text, with Cobra's Head (and its vent) at the left end. Bottom: Proposed Landing site (circled) at lower (left) and higher (right) resolution. Note Landing site is adjacent to large vent on Cobra's Head. (Photos courtesy of NASA)

Schroter's Valley. Despite having Aristarchus Crater impact ejecta, the site was considered 'safe enough' for an experienced pilot.

Apollo 19 (Fig. 8.11): Marius Hills are unique and of complex volcanic terrain, with more volcanic features of every type and representing older and younger episodes of volcanism, ranging from cinder cones to domes to rilles, than any other area of moon. The landing site most extensively studied and favored by Greeley is at the intersection of four domes, within driving distance to a large lava tube, with a skylight.

Apollo 20 (Fig. 8.12): Copernicus is the youngest and brightest lunar crater that is well preserved, with massive central peaks, and an extensive system very visible from earth. A low altitude Lunar Orbiter flight over the rim produced a dramatic view. Boulders tumbled downslope off central peak are larger than the largest rock on earth ('Giant Rock' in California). In an impact feature this large, the central peak and surrounding boulders represent

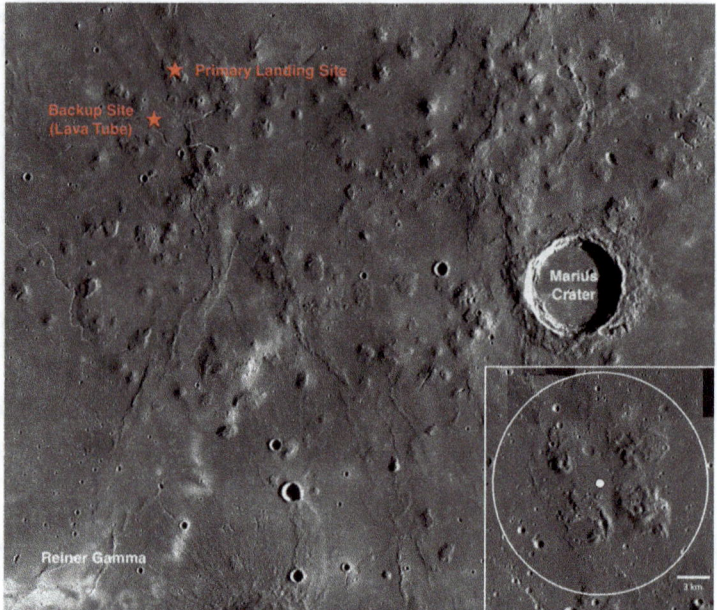

Fig. 8.11 Proposed Apollo 19 Proposed Landing Site Photos. Note Marius Hills extensive and varied volcanic feature and the presence of a magnetic swirl anomaly as described in text. The insert in the bottom right is a close up of the proposed landing site surrounded by four domes plotted in red on the larger map. (Photos courtesy of NASA)

material dredged up from deep within, presumably at the mantle interface which would include the minerals olivine and anorthosite spinel observed remotely on larger basin central peaks elsewhere and could be sampled here. The hard part was finding a 'safe' landing site amid the extensive hummocks and cracks.

8.10 Next Generation Missions

Through the years, scientific interest in unvisited lunar features continued and, from time to time, when this interest coincided with expressed political goals to return to the Moon by a current administration, working groups (e.g., [46, 47]) created lunar surface exploration scenarios to access sites of intriguing geology and in situ resource potential. When such science working groups were convened periodically to revisit landing site selection and exploration (e.g., during the Project Constellation era 2006–2010), their recommendations remained largely consistent with those of the Apollo era. Table 8.1

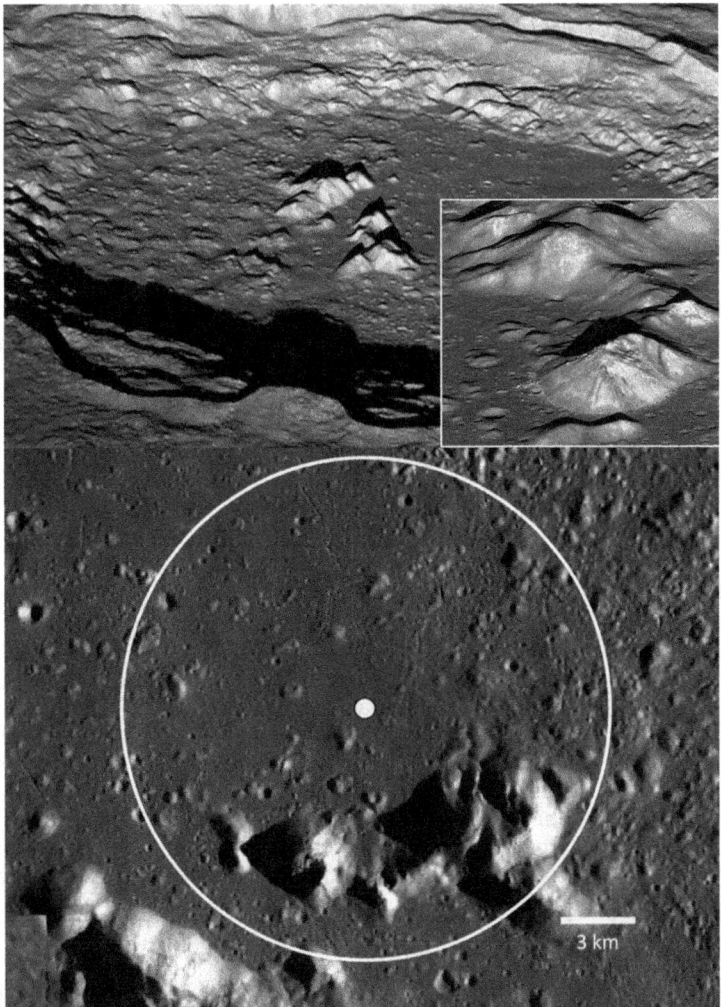

Fig. 8.12 Proposed Apollo 20 Landing Site Photos. Top left is recent LRO LROC view of the Copernicus Crater area as described in the text. The insert to the right is a closeup of the central peaks where boulders downslope of the central peak can clearly be seen. Bottom is the closeup of the proposed landing site just north of the crater's central peak. (Photos courtesy of NASA)

is a representative 'top ten' landing sites from this work. Typically, the Apollo level of capability and technology as the baseline for 'sortie' missions to high priority targets which would have looked very similar to the Apollo J landings, but of somewhat longer duration and with more compact tools for analysis during and between EVAs. Some of the items could be considered a

Table 8.1 Typical 'top ten' future lunar landing sites

Proposed site	Description
Aristarchus Plateau	Potential Field base camp, complex area with potential resources and shelter (lava tubes), representing major volcanic episodes, extensive volcanic valleys, overlaying uplifted basin, recent impact structure
Marius Hills/ Oceanus Procellarum	Potential field base camp, complex area with potential resources and shelter (lava tubes), complex volcanics, every style and major episode of lunar volcanism, most clearly definable domes and cones on Moon
Rima Bode	Extremely dark, with pyroclastics resources. Fresh crater w/dark, thick ejecta blanket. Study pyroclastics, dark mantle material
Orientale Basin	Young basin with well-defined ring structure and minimal mare fill. Study complex ejecta and impact structures radial and concentric to basin
Tycho	Large young impact crater with well-preserved impact structures including prominent rays visible from Earth. Study impact structures
Mare Smythii	Old farside basin patchily covered with some of youngest mare basalt. Study basin formation and late mare volcanism
Central Farside Highlands/ Hertzsprung	Large, older impact basin of central rugged farside feldspathic highlands terrane. Study highlands terrane, underlying farside structure
Tsiolkovsky	Anomalous small, isolated mare flooded farside basin. Study farside basin materials and volcanics
Mare Ingeniii	Prenectarian magnetic swirl anomalies anatipodal to Mare Imbrium in one of few lunar mare features on lunar farside. Study magnetic swirl anomalies
South Pole/Aitken Basin	Potential Field base camp. Complex farside highlands and largest, oldest (Aitken) basin terrain with potential resources (cold traps for ice in largely shaded areas). possible volcanics exposed mantle

'super-sortie': a field base camp requiring more infrastructure to support an outpost for multiple sequential longer crew stays.

NASA space-exploration missions represent the interests of a wide variety of stakeholders. In particular two essential yet conflicting sets of goals need to be reconciled: to spend as much time as possible collecting measurements at sites of interest as soon as possible, and to minimize risk and maximize cost-effectiveness. In developing the 'next generation' scenarios described here, we used JPL's HURON, an advanced modeling tool that uses radial basis functions, kriging, and regression to cover the most essential variations in the complex trade space of a regional-scale mission [53, 54].

This tool is designed to bring the two sets of goals together to create a weighted cost/benefit for a mission by analyzing what the achievement of the

Table 8.2 Apollo-style sorties equipment done currently for 2 groups of 2 astronauts

Equipment	Mass (kg)
Apollo Lunar Rovers (unpressurized, compact)	2 rovers@250 kg
Video and Still Digital Cameras, Voice Communication	6@5 kg = 30 kg
Orientation (location and direction to within meters) without compass and reliance on line of sight astro transponder; Interactive planning/mapping software helmet or handheld	6@1 kg = 6 kg
GeoSurvey tools, inclinometer, ranging	2 rovers@1 kg = 2 kg
Geological Toolkit (rock hammers, chisels, rakes, scales, extendable adjustable scoops, tongs, shovels, carriers	2 rovers@25 kg = 50 kg
Geochemical Toolkit: handheld/rover mounted multi-functional spectrometers (characterize and down select)	6@5 kg = 30 kg
Geotechnical Toolkit: handheld/ rover wheels electrometers, magnetic susceptibility meters	2 rovers@2 kg = 4 kg for trip
Ground Penetrating Radar, rover mounted	2 rovers@5 kg = 10 kg
Magnetometer, rover-mounted or removable	2 rovers@10 kg = 20 kg
Active Seismic Sounder	2 rovers@6 kg = 12 kg
Regolith Drill (15 kg) and 10 Stems (10 kg)	2 rovers@20 kg = 40 kg
Shallow Cores (4 cm drive tubes) (20*0.5 kg each)	2 rovers@10 kg = 20 kg
Sampling supply kit: 2 boxes for 10 kg, 100 bags for 5 kg)	2 rovers@15 kg = 30 kg
Estimated Samples Collected (assuming geochemical toolkit for down select)	200 kg

required science goals will require in the way of engineering design and operational support. In particular we considered:

- key productivity measured from Apollo astronaut operations (Table 8.2)
- study objectives for selected landing site
- scenario physical feasibility as a function of terrains

We looked at a variety of mission scenarios. Prioritization of activities and requirements for resources, including time and equipment for those activities, were provided by Apollo studies discussed above. Initially, the largest constraint was found to be initial limit of 4 EVA hours per day imposed by the use of a pressurized rover for transportation, resulting in a 90-day mission. The study assumed the astronauts were traveling in two pressurized rovers accompanied by a slower robotic vehicle to carry equipment. The pressurized vehicle was not designed to allow sample collection and terrain characterization when the astronauts were inside. Subsequently, we relaxed this constraint, allowing astronauts to use unpressurized vehicles in spacesuits during EVAs. Travel in their spacesuits in unpressurized vehicles, with direct access to sample collection and terrain characterization at all times, accompanied by a larger and slower robotic vehicle towing a rechargeable life support system they would use as a 'camper' between EVAs reduced the EVA time constraint and thus the time required for the mission even more. Constraints considered included EVA time, EVA time/locality, rover speed, number of primary and secondary localities included, and air lock egress/ingress time. Outputs for 5 parameters were considered including science value, productivity, cost, mission duration, and kilometers traversed. Science value was enhanced by increasing EVA time.

The work described here, originally published in 2010 [46, 47] involved creating operational scenarios at two levels. Most of these operational scenarios were seen as follow-ons to Apollo J missions (15, 16, and 17) and baselined Apollo-era hardware, including rovers and toolkits, but also included comparable versions of state-of-the-art portable instrumentation, such as a compact ground-penetrating radar (Table 8.2). Scenarios also assumed the landing of two 2-person Lunar Modules, or a somewhat larger 4-person Lunar Module, two lunar rovers and toolkits, and 1 week on the surface and four EVAs per pair of astronauts. The second group, regional studies, utilized additional surface transportation hardware capability and assumed at least 4 astronauts. The process of target selection is described above. The current availability of high-quality topography and surface roughness data, from LRO or Lunar Orbiter data enhanced through the Lunar Orbiter Image Recovery Project (LOIRP), greatly enhances the capability to plan operational scenarios for field work on the lunar surface. For detailed route planning, data with resolution on the scale of a meter or less is crucial.

Sortie Studies: Although sortie-level operational scenarios were created for Marius Hills, Nectaris Basin and Olivine Hill (Table 8.3), the scenarios created for Marius Hills is described here in detail [46] (Figs. 8.13, 8.14, 8.15, and 8.16): Typically, "…as in the Apollo program, we used the rules that pairs of astronauts would not be allowed to go more than 5 kilometers, walk back distance, from the lander." Each crew did 4 EVAS (on days 2, 3, 5, and 6). We considered larger diameters sites as well (20 and 30 km) (Fig. 8.17). In one case "the astronauts used two rovers, with one parked 5 km from the lander, to go 10 km from the lander for a particularly interesting site, so we stuck with the original 'safer' architecture with 5 km radius. Geophysical experiments would be deployed near the lander during EVA 1, as they were for the Apollo J missions. Marius Hills (Figs. 8.13 and 8.14) brackets the full range of lunar mare volcanism styles and ages, including the two major mare formations (Flamsteed and Marius). It includes the largest variety of volcanic features found in any are of the Moon, including shields, domes, cones. Marius Hills has several prominent rilles and is adjacent to Reiner Gamma, a magnetic swirl anomaly. We identified two representative landing sites in the

Table 8.3 Sortie-level operational scenarios

Sortie	Site description, 10 km radius, 1 week, 2 UPRs, 4 astronauts	Alternative
Marius Hills	Brackets full range of lunar mare volcanism styles and ages, including two major mare formations of different ages (Flamsteed and Marius) Samples largest variety of volcanic features/deposits including shields, domes, cones, rilles Samples full range basin-related tectonic features including wrinkle ridges, fractures, faults	Regional scale study originating from Aristarchus
Olivine Hill	Anomalous olivine rich deposit centrally located near exposed basin floor, iron anomaly in SPA Provide clues to origin largest, oldest confirmed basin age, crustal composition, origin Crossed by well-defined ray to reveal composition of surrounding highlands	Regional scale study originating in northern SPA
Nectaris Basin Edge	Establishes Absolute Basis for Lunar Chronology by determining Nectaris Basin Age Establishes age/structural relationship between basin formation and volcanic activity Samples portions of exposed highland basin massif, basin floor, and mare basalt fill	Adequate considering relatively narrow focus

southern Marius Hills (55 West, 13 North), we called Marius West and Marius East (Figs. 8.15, 8.16, and 8.17). Marius East field work involved characterizing low titanium and high titanium volcanic structures, volcanic vent complexes, wrinkle ridge complexes, fresh craters, and a collapsed crater chain. All four astronauts made the trip to Reiner Gamma to do the first in situ characterization of the largest lunar magnetic anomaly. Marius East field work involved characterizing both sides of an extensive volcanic rille, as well as high-titanium and low-titanium dome and vent complexes."

Regional studies: Beyond the globally distributed separate sortie approach initially favored by the scientific community, an outpost model allows for in-depth contextual coverage, progressively greater stay duration and ultimately permanent occupation. In our regional scale studies [47], we looked at science activities and requirements for exploration on 100–1000 km scales, beyond the scale achievable at an outpost early on or at a single sortie landing site alone. These longer, larger scale efforts would require longer range and duration mobility capability. Rovers would not have to be constantly pressurized to carry human crew (with larger mass and power demands for life support and less available for mobility) from an outpost. Long range vehicles could

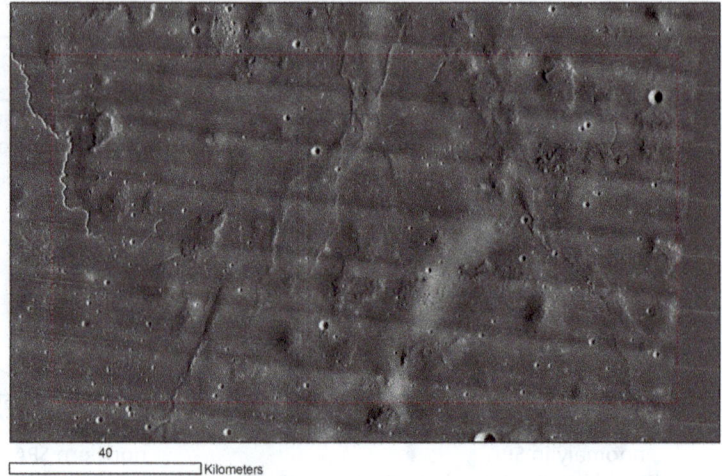

Fig. 8.13 Lunar Orbiter Photo of Marius Hills showing diversity of features including volcanic cones and domes, mare Flows, sinuous rilles, straight rilles (faults?), wrinkle ridges, impact craters, secondary crater chains, and magnetic swirl anomaly Reiner Gamma (bottom center). (Photo Courtesy of NASA)

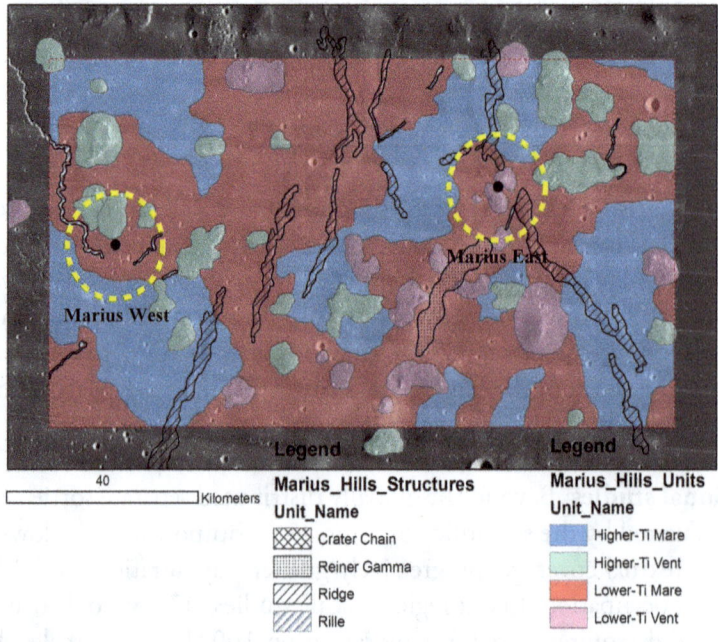

Fig. 8.14 Geological map derived from photo seen in Fig. 13 with rilles, ridges, crater chains, and Reiner Gamma marked. 10 km radius circles indicate two potential landing sites (Marius West and Marius East) identified in study described in text [46]. (Courtesy of Clark et al., Fig. 5, 2010, GSA Bulletin)

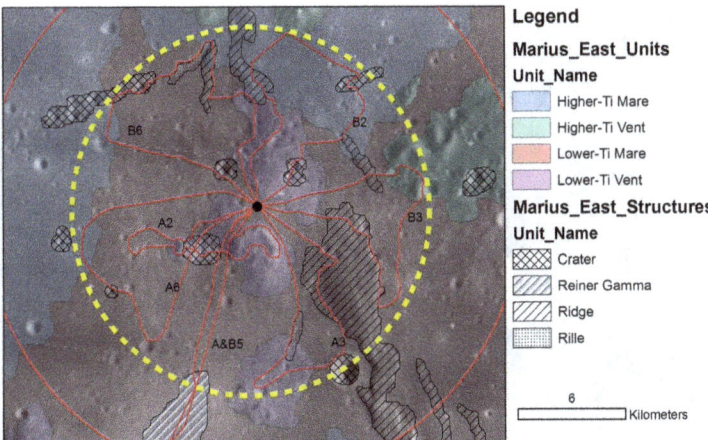

Fig. 8.15 Marius East landing site with traverses, as described in text, marked in red. Visits planned to all identified morphologic and spectral units, enables analysis of unit contacts and organization of local developmental history including insight into impact, tectonic, volcanic, regolith, and magnetic processes [46]. (Courtesy of Clark et al., Fig. 8, 2010, GSA Bulletin)

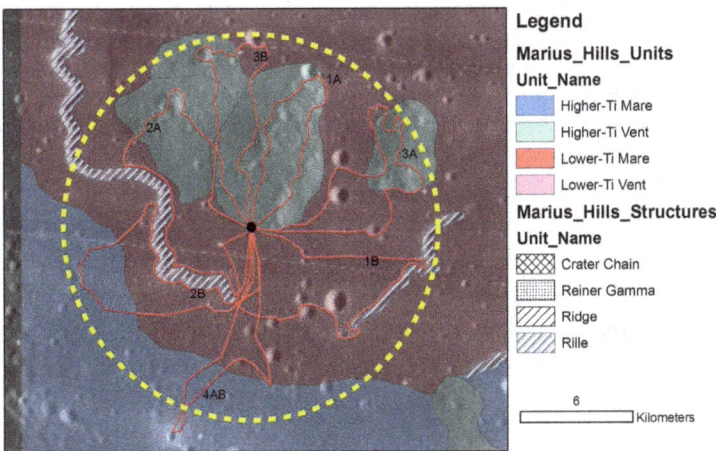

Fig. 8.16 Marius West landing site with traverses, as described in text, marked in red. Visits planned to all identified morphologic and spectral units, enables analysis of unit contacts and organization of local sequential history, with visits to both sides of a volcanic rille and magnetic swirl anomaly, providing insight into impact, volcanic, and regolith processes [46]. (Courtesy of Clark et al., Fig. 9, 2010, GSA Bulletin)

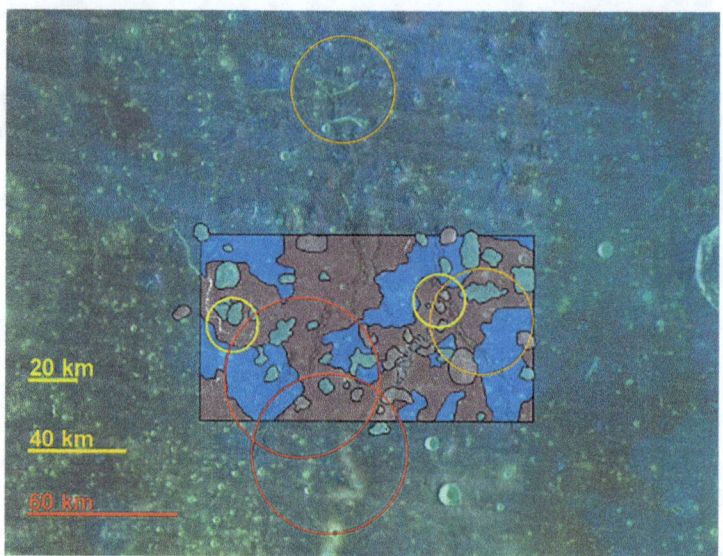

Fig. 8.17 Landing Site Selection as a function of study area size [46] (Fig. 6). Sites were selected on the basis of providing the best access to the widest range of volcanic and tectonic features representing the formation and history of Marius Hills in the context of the western nearside. Within study areas of given radii 10, 20, and 30 km, respectively. Two 10 km radius (20 km dimeter) and two 20 km radius sites met most of those criteria. Each 30 km radius site met those criteria, which included the magnetic swirl anomaly, but with different emphases. This study focused on the 10 km radius sites because they best fit the 'sortie' architecture under consideration [46]. (Courtesy of Clark et al., 2010, Fig. 6, GSA Bulletin)

perform telerobotic exploration and carry equipment, including rechargeable pressurization chambers, for suited astronauts landed at the sortie site or traveling there from the outpost in rechargeable unpressurized vehicles and then exploring an area over days or weeks before returning to an outpost. The strategy and tactics for planning and choosing temporary 'field camps' and logistical requirements to support such work become essential architecture.

Contextual or regional scale field studies could significantly enhance what is learned locally, while greatly enhancing insight into major science questions using carefully selected sites. Table 8.4 indicates what additional equipment (beyond that used for earlier sorties) we recommended to support regional studies. These scenarios could be thought of as consecutive sorties on a regional scale. The minimum of personnel, equipment, and activity would be 4 astronauts, a shared portable life support system, and two fully equipped (as in the lunar rover) unpressurized rovers, performing 4–8 h EVAs, to perform at least one major site visit a day. Samples could be returned with the astronauts or remain stored in the portable lab/life support system.

Table 8.4 Regional scale studies equipment beyond sorties in Table 8.2

Equipment (Could be in Rechargeable Lab/Life Support System for real time analysis or at supplied by and returned to Outpost for later analysis)	Mass (kg) (to support field work beyond that done in sorties)
Rechargeable 'Morph Lab' Lab/Life Support System (delivered in advance)	4000 kg
Robotic Assistant for well-defined repetitive tasks (systematic sample collection, encapsulation, preparation, curation in designated area of interest)	2@100 kg
Laboratory (Petrographic microscope, spectrometers, meters; polishing, slicing, thin sectioning equipment; particle size/shape analyzer)	2@75 kg
Estimated Samples Collected	30 kg/EVA

Table 8.5 Regional scale operational scenarios

Scenario	Activity	Options
SPA/Apollo Basin (North)	Three 2 week trips of up to 1000 km from outpost south of Apollo Basin	Could also do as three 1 week sorties with landing sites at Olivine Hill, Oppenheimer Pyroclastics, and Apollo Basin
SPA/Antipode (West)	One 2 week trip of up to 1000 km from outpost in Mare Ingenii	Coul also o as two 1 week sorties with landing sites in Mare Ingenii and Antipodal terrain
SPA/South Pole (South)	One 2 week trip of up to 1000 km from landing site near South Pole	Trip from South Pole in other direction to Malapert/Leichert Mountain complex would focus on SPA Basin formation
Tsiolkovsky	Three 2 week trips of up to 1000 km from outpost on western edge of Tsiolkovsky	Could also do as six shorter sorties originating from outpost near central peak for Tsiolkovsky with less exploration of surrounding basin
Aristarchus Plateau	Four 2 week trips of up to 1000 km from outpost just west of Plateau	Could do plateau as series of four 1 week sorties from plateau outpost with partial circumnavigation and then do separate 1 week sorties with outpost Marius Hills, Lichtenberg Crater, or other features of interest

As in the case of the sorties, the three selected targets were those that have typically remained in high priority regional site lists since the Apollo program (Table 8.5). They included South Pole Aitken Basin, farside Tsiolkovsky-Stark, and nearside Aristarchus Plateau. South Pole Aitken Basin (Figs. 8.18, 8.19, 8.20, and 8.21) and Tsiolkovsky-Stark (Figs. 8.22, 8.23, and 8.24) will be looked at here in detail.

Fig. 8.18 Tsiolkovsky Basin of the lunar farside as seen from Lunar Orbiter. (Photo courtesy of NASA)

Tsiolkovsky is a distinctive, farside feature, compact mare-filled, and set in the middle of the ancient pre-Nectarian Tsiolkovsky-Stark Basin. Quoting from [47] "The area is anomalous due to its depth of mare fill in a relatively 'thick' portion of farside crust and its age (slightly older than Imbrium), typical of nearside mare terrane rather than farside highland crater, yet has extremely well-defined impact structures, including central peak, and perhaps sits on SPA outer ring. The central peak is displaced to the north and west, and the crater apparently has greater ejecta deposition on the south wall, and more extensive impact melt pools in the NW and SW, indicating asymmetry…" in the impact event and "…basin structure. The crater exhibits complex slumping along concentric faults in the rim terraces, and evidence for landslides can be seen at the bottom of the terraces. The central peak exhibits extensive mass wasting, with visible large blocks, and young age, due to its steep slopes, with the exception of the relatively flat plateau, which exhibits near saturation cratering and the oldest age. As in other large impact structures on the farside, the central peak, originating from the deepest part of the

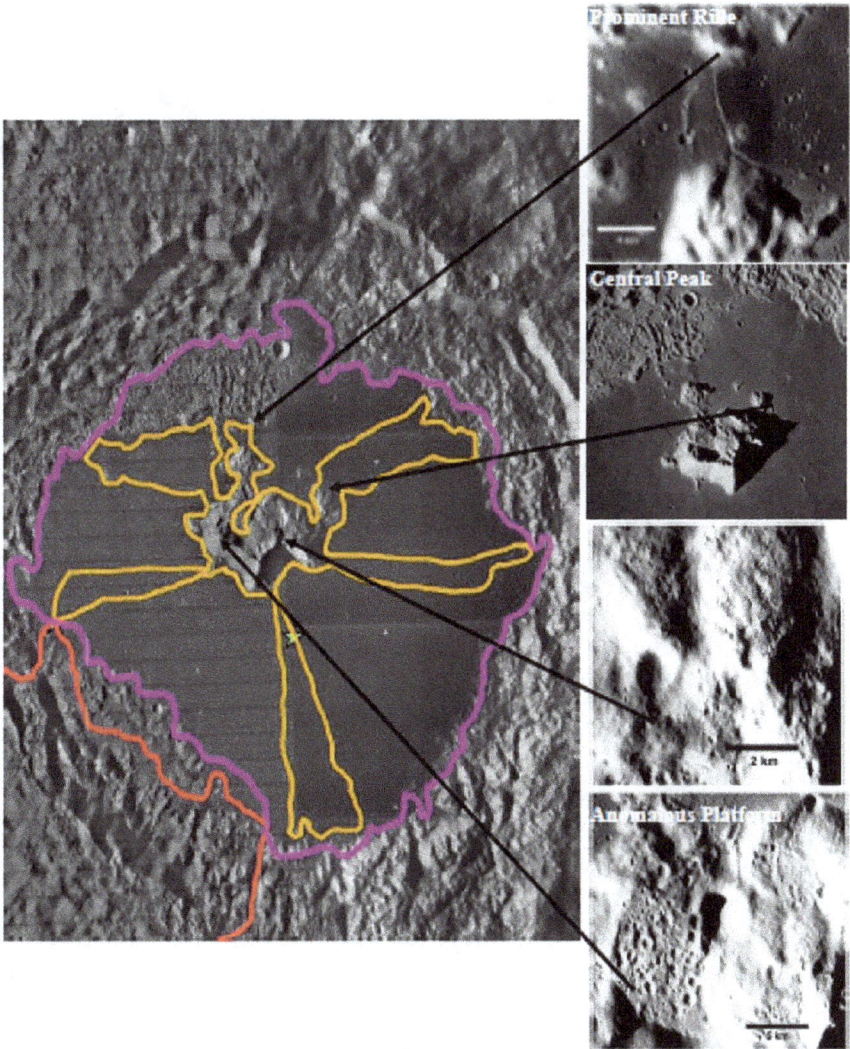

Fig. 8.19 Tsiolkovsky Basin regional scale scenario with selected peak structural study features for central peak and interior routes as described in the text [47]. (Courtesy of Clark et al., Fig. 8a, 2010, GSA Bulletin)

crust, is apparently troctolitic, with the walls appearing more noritic. Mare fill has NS compositional dichotomy. Crater statistics indicate at least two populations, potentially representing two episodes of mare volcanism, on the mare floor, although, up to 8 flooding episodes are theoretically possible, based on the crater's depth."

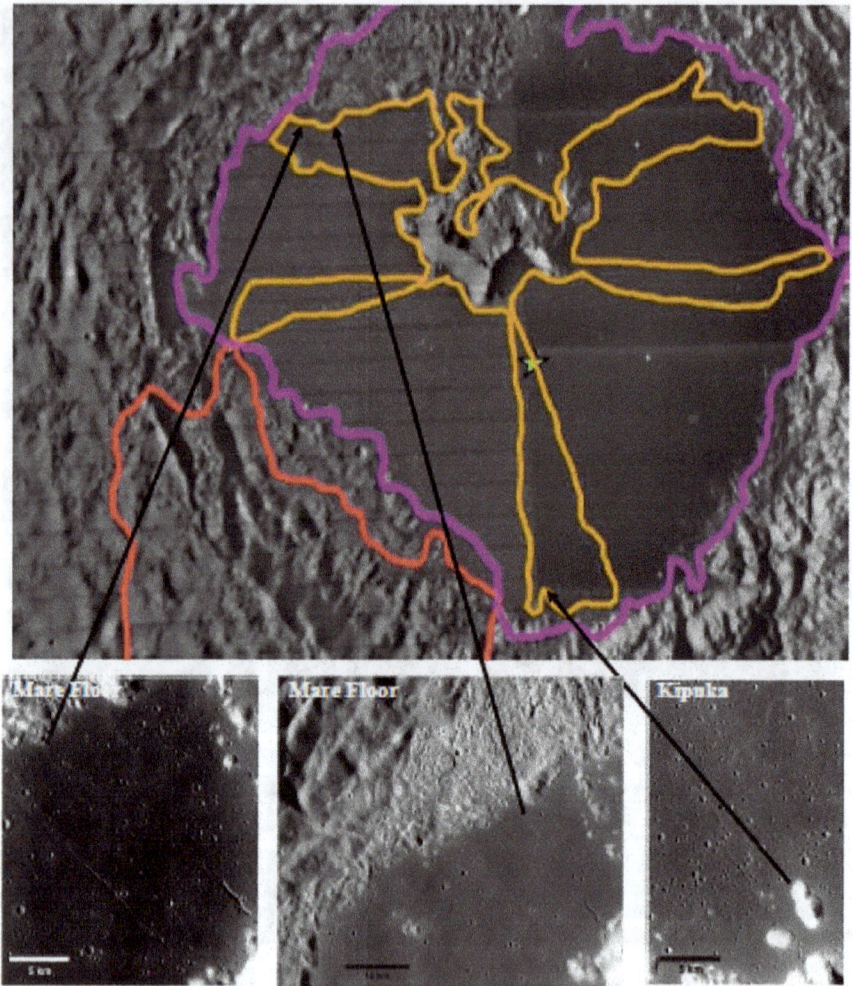

Fig. 8.20 Tsiolkovsky Basin regional scale scenario with selected mare basin and rim study features as described in the text [47]. (Courtesy of Clark et al., Fig. 8b, 2010, GSA Bulletin)

Field work would address a) structure, composition, and formation of its impact features, including its extremely well defined central peak, complex ring terraces and exposed impact melt floor, b) its numerous ridges, scarps, and valleys oriented radially and concentrically oriented, c) the nature of highland volcanism, including evidence for mare flow fronts representing episodes of mare volcanism, such as the dark patch north of the central peak, and potential vents surrounding the central peak, and d) formation and bombardment history of the surrounding basin, and its relationship to SPA. A series of routes would focus on

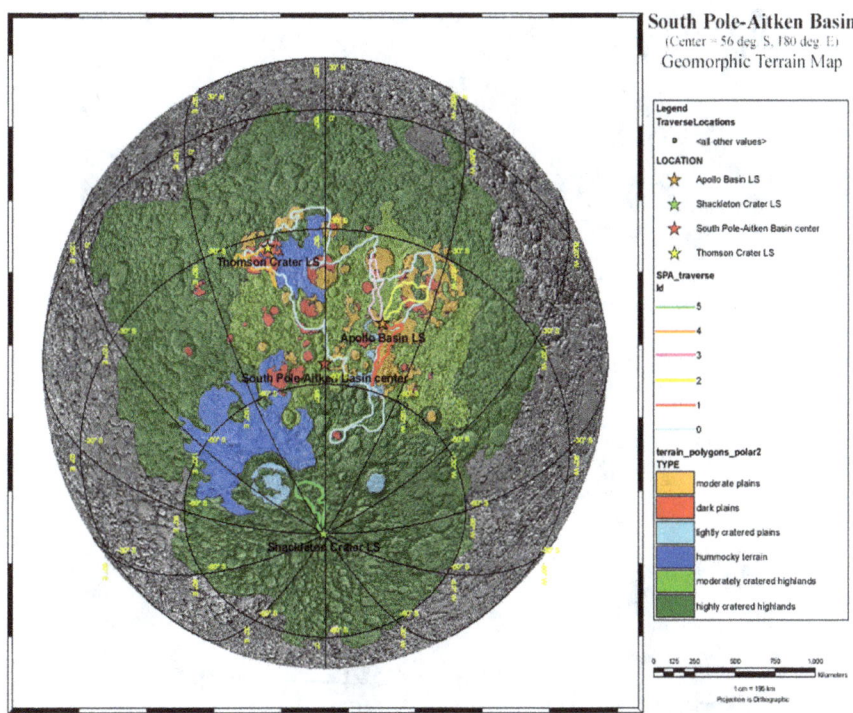

Fig. 8.21 Geological unit map of the Lunar Farside South Pole Aitken Basin with potential traverse routes superimposed. Note variations in major terrains, with less cratered terrain within the basin and dark plains implying cryptomare distributed largely in the northern portion. Landmarks discussed in text are marked with stars [47]. (Courtesy of Clark et al., Fig. 4a, 2010, GSA Bulletin)

the central peak area, traverses of the basin floor, circumnavigation to study the surrounding walls, uncovered floor, and impact ejecta, and crypto-mare in the surrounding ancient basin. Features of particular interest include the asymmetrical central peak, and a prominent northern rille and dark patch associated with it, networks of radial and concentric around the central peak and rim, exposed impact melt and landslides on the western rim, Tsiolkovsky's ejecta partially obliterating Fermi crater to the west and south, and Waterman crater cryptomare to the south.

South Pole-Aitken Basin (SPA) is the, quoting from [47] "largest and oldest confirmed basin on the Moon, with multi-ring structure extending over a large area, perhaps as far as Mare Moscoviense in the north and Malapert Mountain in the south."

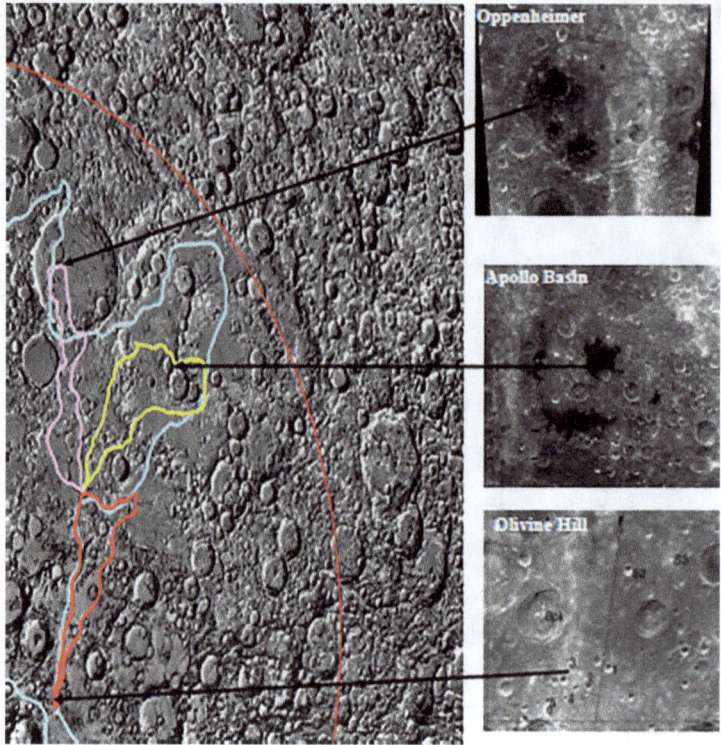

Fig. 8.22 Regional scale mission traverse routes from Apollo Basin outpost, with close-ups of feature of primary interest, Apollo Basin, Olivine Hill, and Oppenheimer pyroclastics [47]. (Courtesy Clark et al., Fig. 7a, GSA Bulletin)

Its formation caused crustal thinning, particularly in the northern half of SPA, exposing deeper crust and even penetrating into the mantle potentially explaining hemispheric compositional asymmetries. Thorium shows a increase and iron a decrease to the west, potentially indicating greater presence of KREEP rich basalt. Basin formation models fail to predict the dearth of mare basalt flows in SPA. Northern half of SPA contains volcanic pools implying asymmetry in basin structure. Largely local floor material should be exposed near the center of basin, near Olivine Hill anomaly. Jackson Ray which extends across the northern half of the basin would also give indications of the character of the surrounding farside highlands. Geochemical trends imply heterogeneity in crust. Antipodal anomalies include magnetic swirls of Mare Ingenii, 'mound and groove' terrain west of Mare Ingenii. The origin of high Th anomalies there, antipodal deposition or local, is still debated. Access to the South Pole would yield an opportunity to explore permanently shadowed craters for volatiles.

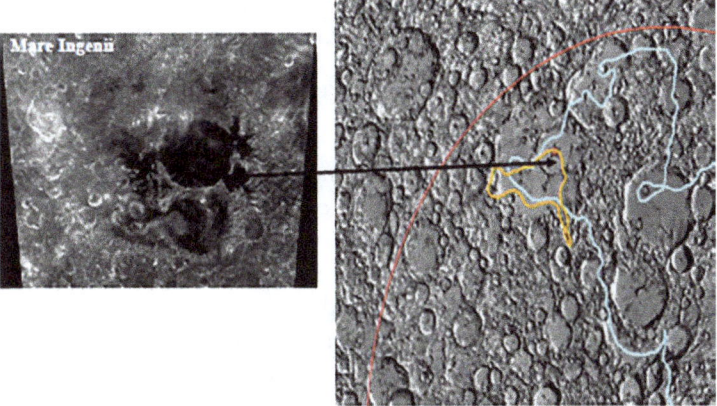

Fig. 8.23 Regional scale mission traverse routes from Thomson Crater outpost, with close-up of feature of primary interest, Mare Ingenii magnetic swirl anomaly [47]. (Courtesy of Clark et al., Fig. 7b, 2010, GSA Bulletin)

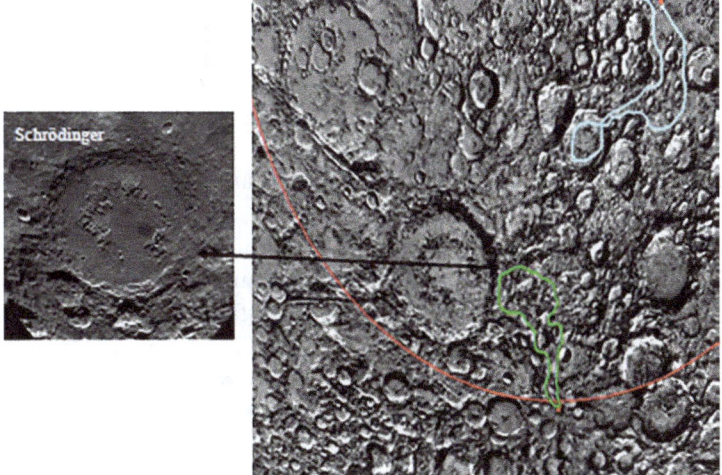

Fig. 8.24 Regional scale mission traverse routes from Shackleton Crater outpost, with close-up of feature of interest Schrodinger Basin [47]. (Courtesy of Clark et al., Fig. 7c, 2010, GSA Bulletin)

Field work would involve characterizing basin structure, farside highlands, and volcanic terrane structure through the use of radial traverses with ground penetrating radar, magnetometer, and sample collection in the northern half of SPA, including the antipodal magnetic features of Mare Ingenii and the surrounding 'mound and groove' terrane. Toward the center of the basin, near Olivine Hill, or further south, where volcanic cover is lacking, the floor materials themselves could be sampled. From a South Pole site, Malapert Mountain could be surveyed as an outer ring rampart and permanently shadowed craters could be surveyed for volatiles.

A regional scale exploration would consist of a series of trips, traversing from the northern central area north, northwest, and south/southwest toward the center of the basin, to cover the features of particular interest. A similar traverse from the southern edge of SPA would complete our understanding of the underlying basin structure and the north south asymmetry of SPA. Prominent sites would include Olivine Hill, an olivine-rich feldspar (troctolite) anomaly near Bose Crater in the middle of SPA, Apollo Basin with dark mare deposits in northeast and ancient mare deposits in the southwest), Oppenheimer crater with pyroclastic deposits in the northwest and a prominent ray in the east, Mare Ingenii, with antipodal terrain and a magnetic swirl anomaly, and high thorium anomalies to the east and southeast in northwest SPA, and Schrodinger Basin, the most prominent basin in southern SPA, and Malapert Mountain, representing an outer ring of SPA.

References

1. Apollo Missions, LPI, https://lpi.usra.edu/lunar/missions/apollo/. visited 10/1/2024
2. Donovan, J., Shoot for the Moon. Little Brown and Company, 464 p., 2019.
3. Kranz, G., Failure is not an option, Simon &Schuster, 416 p., 2009.
4. Apollo 11 Preliminary Science Report NASA SP-214. 1970. https://ntrs.nasa.gov/api/citations/19700025955/downloads/19700025955.pdf. visited 10/1/24.
5. Tylko, J., MIT and navigating the path to the Moon, AeroAstromagazine, MIT, 2009. https://web.mit.edu/aeroastro/news/magazine/aeroastro6/mit-apollo.html. visited 10/1/2024.
6. Harland, D., Apollo 12- On the Ocean of Storms, Praxis Books, 530 p., 2011.
7. Apollo 12 Preliminary Science Report NASA SP-235. 1970. https://ntrs.nasa.gov/api/citations/19700025955/downloads/19700025955.pdf. visited 10/1/24.
8. Mitchell, E., E. Mahoney, B. Cox, Earthrise: My adventures as an Apollo 14 Astronaut, Chicago Review Press, 192 p., 2014.
9. Godwin, R., Apollo 14: The NASA Mission Reports, Apogee Book Series, 14, 332 p. 2001.
10. Apollo 14 Preliminary Science Report NASA SP-272. 1971. https://www.nasa.gov/wp-content/uploads/static/history/alsj/a14/as14psr.pdf. visited 10/1/24.

11. Clark, P.E., Revolution in Field Science: Apollo Approach to Inaccessible Surface Exploration, Earth, Moon, and Planets, 106, 2–4, 133–157, 2010. http://www.springerlink.com/content/uk63t1n6426016l7/
12. Apollo 15 Preliminary Science Report NASA SP-289. 1972. https://www.nasa.gov/wp-content/uploads/static/history/alsj/a15/as15psr.pdf. visited 10/1/24
13. Apollo 16 Preliminary Science Report NASA SP-315. 1972. https://www.nasa.gov/wp-content/uploads/static/history/alsj/a16/as16psr.pdf. visited 10/1/24
14. Apollo 17 Preliminary Science Report NASA SP-330. 1973. https://ntrs.nasa.gov/api/citations/19740010315/downloads/19740010315.pdf. visited 10/1/24
15. NASA/MSC Lunar Missions Office, J–1 Mission Science Requirements, MSC–02536. 1970.
16. NASA Manned Spaceflight Office, Apollo 15 Mission Operation Report, NASA M–933–71–15, 1971.
17. Allen, J., Summary of Scientific Results, Apollo 15 Preliminary Science Report, NASA SP–289, 2-1–2-11, http://www.hq.nasa.gov/alsj/a15/a15psr.html, 1972.
18. Swann, G., N. Bailey, R. Batson, V. Freeman, M. Hait, J. Head, H. Holt, K. Howard, J. Irwin, G. Schaber, D. Scott, L. Silver, R. Sutton, G. Ulrich, H. Wilshire, E. Wolfe, Preliminary geologic investigations of the Apollo 15 landing site, Apollo 15 Preliminary Science Report, NASA SP–289, 5-1 – 5-112, http://www.hq.nasa.gov/alsj/a15/a15psr.html, 1972.
19. Spudis, P. and G. Ryder, Geology and petrology of the Apollo 15 landing site: Past, present, and future understanding, Eos Trans AGU, 66, 721–726, 1985.
20. Vaniman, D., R. Reedy, G. Heiken, G. Olhoeft, W. Mendell, The Lunar Environment, in The Lunar Sourcebook, G. Heiken, D. Vaniman, B. French, Eds., 27–600, 1991.
21. McKay, D., G. Heiken, A. Basu, G. Blanford, S. Simon, R. Reedy, B. French, J. Papike, The Lunar Regolith, in The Lunar Sourcebook, G. Heiken, D. Vaniman, B. French, Eds., Cambridge University Press, 285–356, 1991.
22. Spudis, P. and C. Pieters, Global and Regional Data about the Moon, in The Lunar Sourcebook, G. Heiken, D. Vaniman, B. French, Eds., Cambridge University Press, 595–632, 1991.
23. Apollo 15 Field Geology Investigation Team, Apollo 15 Lunar Samples – Preliminary Description, Science, 175, 4020, 363–375, 1972.
24. Apollo 15 Preliminary Examination Team, Geologic Setting of Apollo 15 Samples, Science, 175, 4020, 407–415, 1972.
25. Basu, A., D. McKay, R. Fruland, Origin and modal petrography of Luna 24 soils. Mare Crisium: The View from Luna 24, R. Merrill and J. Papike, Eds, Pergamon, New York, 321–337, 1978.
26. NASA/MSC Science Mission Support Division, J–2 Mission Science Planning Document, MSC–04143, 1972.
27. Hinners, N., Apollo 16 Site Selection, Apollo 16 Preliminary Science Report, NASA SP–315, 1-1–1-3, http://www.hq.nasa.gov/alsj/a16/a16psr.html, 1972.

28. England, A., Summary of Scientific Results, Apollo 16 Preliminary Science Report, NASA SP–315, 3-1-3-8, http://www.hq.nasa.gov/alsj/a16/a16psr.html, 1972.
29. Muehlberger, W., R. Batson, E. Boudette, C. Duke, R. Eggleton, D. Elston, A.England, V. Freeman, M. Hait, T. Hall, J. Head, C. Hodges, H. Holt, E. Jackson, J. Jordan, K. Larson, D. Milton, V. Reed, J. Rennilson, G. Schaber, J. Schafer, L. Silver, D. Stuart–Alexander, R. Sutton, G. Swann, R. Tyner, G. Ulrich, H. Wilshire, E. Wolfe, J. Young, Preliminary geologic investigation of the Apollo 16 landing site, Apollo 16 Preliminary Science report, NASA SP–315, 6–1 – 6–81, http://www.hq.nasa.gov/alsj/a16/a16psr.html, 1972.
30. James, O., Petrologic and age relations in Apollo 16 rocks: Implications for subsurface geology and the age of the Nectaris basin. Proc Lun Plan Sci 12B, 209–233, 1981
31. Spudis, P., Apollo 16 site geology and impact melts: Implications for the geologic history of the lunar highlands, Proc Lun Plan Sci Conf 15th, JGR, 91, C95–C107, 1984.
32. Stoffler, D., A. Bischoff, R. Borchardt, A. Burghele, A. Deutsch, E. Jessberger, R. Ostertag, H. Palme, B. Spettel, W. Reimold, K. Wacker, H. Wanke, Composition and evolution of the lunar crust I the Descartes Highlands, Apollo 16. Proc Lun Plan Sci Conf 15th, JGR, 89, C449–C506, 1985.
33. Ulrich, G., C. Hodges, W. Muehlberger, Geology of the Apollo 16 area, Central Lunar Highlands. USGS Prof Paper 1048, 1–539, 1981.
34. Vaniman, D. and J. Papike, Lunar highland melt rocks: Chemistry, petrology, and silicate mineralogy, Proceedings of the Conference on the Lunar Highlands Crust, J. Papike and R. Merrill, Eds., Pergamon, New York, 271–337, 1980.
35. Apollo 16 Field Geology Investigation Team, Apollo 16 Exploration of Descartes – Geologic Summary, Science, 179, 4068, 62–69, 1973.
36. Apollo 16 Preliminary Examination Team, Apollo 16 Lunar Samples – Petrographic and Chemical Description, Science, 179, 4068, 23–24, 1973.
37. NASA/MSC Science Mission Support Division, J–3 Mission Science Planning Document, MSC–05871, 1972.
38. Hinners, N., Apollo 17 Site Selection, Apollo 17 Preliminary Science Report, NASA SP–330, 1-1-1-5, http://www.hq.nasa.gov/alsj/a17/a17psr.html, 1973.
39. Parker, R., Summary of Scientific Results, Apollo 17 Preliminary Science Report, NASA SP–330, 3-1-3-7, http://www.hq.nasa.gov/alsj/a17/a17psr.html, 1973.
40. Muehlberger, W., R. Batson, E. Cernan, V. Freeman, M. Hait, H. Holt, K Howard, E. Jackson, K. Larson, V. Reed, J. Rennilson, H. Schmitt, D. Scott, R. Sutton, D. Stuart–Alexander, G. Swann, N. Trask, Ulrich, H. Wilshire, E. Wolfe, Preliminary geologic investigation of the Apollo 17 landing site, Apollo 17 Preliminary Science report, NASA SP–330, 6-1 – 6-91, http://www.hq.nasa.gov/alsj/a17/a17psr.html, 1973.

41. Mitchell, J, D. Carrier, N. Costes, W. Houston, R. Scott, H. Hovland, Soil Mechanics. Apollo 17 Preliminary Science Report, NASA SP–330, 8-1 – 8-22, http://www.hq.nasa.gov/alsj/a17/a17psr.html, 1973.
42. Schmitt, H., Apollo 17 report on the valley of Taurus–Littrow, Science, 182, 681–690, 1973
43. Schmitt, H. and G. Cernan, A geological investigation of the Taurus–Littrow Valley, Apollo 17 Preliminary Science Report, NASA SP–330, 5-1 – 5-21, http://www.hq.nasa.gov/alsj/a17/a17psr.html, 1973.
44. Apollo 17 Field Geology Investigation Team Geologic Exploration of Taurus–Littrow – Apollo 17 Landing Site, Science, 182, 4113, 672–680, 1973.
45. Apollo 17 Preliminary Examination Team, Apollo 17 Samples – Chemical and Petrographic Description, Science, 182, 4113, 659–672, 1973.
46. P.E. Clark, S. Mest, N. Petro, J. Bleacher. Advanced Regional Scale Scenarios for Lunar Surface Exploration, GSA Bulletin Special Issue, http://specialpapers.gsapubs.org/content/483/547.abstract, 2010.
47. P.E. Clark, S. Mest, N. Petro, J. Bleacher. Plan for a Human Expedition to Marius Hills and its Implications for Viable Surface Exploration Architecture, GSA Bulletin, Special Issue, http://specialpapers.gsapubs.org/content/483/519.abstract, 2010.
48. Hammond, A., Lunar Science: Analyzing the Apollo Legacy, Science, 179, 4080, 1313–1315, 1973.
49. Canceled Apollo Missions, Wikipedia, https://en.wikipedia.org/wiki/canceled_apollo_missions. visited 10/1/24
50. NSSDC, Apollo 18 through 20 - The Cancelled Missions, https://nssdc.gsfc.nasa.gov/planetary/lunar/apollo_18_20.html
51. Longo, A., Lost Moon: Reconstructing the Missions of Apollos 18, 19, and 20 (Part 1). https://www.americaspace.com/2023/08/28/lost-moon-reconstructing-the-missions-of-apollos-18-19-and-20-part-1/
52. Longo, A. Lost Moon: Reconstructing the Missions of Apollos 18, 19, and 20 (Part 2). https://www.americaspace.com/2023/09/05/lost-moon-reconstructing-the-missions-of-apollos-18-19-and-20-part-2/
53. Weisbin, C.R., P.E. Clark, A. Elfes, J.H. Smith, J. Mrozinski, V. Adumitroaie, H. Hua, K. Shelton, W. Lincoln, R. Silberg, M. Lupisella. Reconciling Scientific Aspirations and Engineering Constraints for a Lunar Mission via Hyperdimensional Interpolation, AIAA Space Conference 2010, Volume 4, p. 2648–2652, 2010.
54. Clark, P.E., C.R. Weisbin, K.E. Shelton, J.H. Smith, J. Mozinski, W. Lincoln, A. Elfes, H. Hua, V. Adumitroaie, R. Silberg (2009) Lunar surface mission productivities as a function of mission duration and weighted science objectives, Space 2009, 2009. AIAA–2009–6632–995.pdf. http://www.aiaa.org/agenda.cfm?dateget=all

9

Beyond Apollo: Applying Lessons Learned

9.1 Once and Future Lunar Exploration: Future

Finally applying lessons learned from the Apollo program (and in the intervening years) to establish an ongoing presence on the Moon will truly establish America as a space-faring nation, as originally intended when Kennedy made the commitment. Learning how to operate in that extremely challenging environment long-term will pave the way for in-depth exploration of Mars and asteroids. This will be an opportunity to dramatically improve the state of the art for autonomous systems, particularly their efficient (minimum resource) use for autonomic functions (environment/actuator interface) for real-tome reconfigurable and manageable housing, transportation, life support, power production, manufacturing, environmental monitoring, resource extraction and management. In an environment with the built in fail/safe of near real time, we will experiment with developing the most effective human/machine interface for more autonomous operation required for Mars, transition from automation, to teleoperation, to autonomous control with manual override for essential systems. The pace at which these activities unfold, and an economically sustainable model emerges, as it has in low Earth orbit, will depend on political will. As in the case of the Apollo program, there are no technological challenges so great that they can't be overcome.

The focus in this final chapter is on the application of lessons learned as a result of the Apollo Program, lessons discussed or implied in the preceding chapters. How well have we incorporated or neglected these lessons in developing robotic missions and planning human exploration missions over the last five decades, leading up to the current Lunar Initiative? How well do existing and planned space architectures support strategic and economically sustainable advances in extraterrestrial exploration beyond 'expedition level' activity begun in the Apollo era, when the ultimate goal was to send humans beyond cislunar space? In what ways does the cubesat paradigm, a direct descendant of the major microminiaturization thrust of that era, embody and advance that goal?

Those planning and participating in activities to establish a longer-term presence on the lunar surface have faced and will continue to face challenging constraints in the foreseeable future until some degree of sustainability is established for cislunar activities. Constraints include the limited availability of in situ resources and, even after half a century, lack of an established architecture for cislunar operations with or without humans in the loop. These constraints are costly because far more challenging than the logistic barriers that were faced to set up outposts after preliminary exploration of even the remotest areas on Earth as discussed in Chap. 2.

Investment to overcome these challenges has been typically limited during previous decades when visionary leadership, development of breakthrough technologies essential for operating in extreme environments with reduced resources, and willingness to take short term risk in creating an infrastructure for long term gain has been lacking in every sector and not just aerospace. On the other hand, Apollo era programs (yes, including the essential precursors) provided a basis we can build on for 'what works' (and what doesn't) in remote exploration both with and without humans in the loop. Advanced planning with the best available reconnaissance, leading to realistic simulations, intensive training, and efficient tooling, played a critical role for the Apollo program. The development of systematic, largely hands-free methods of verbal description, with in situ photography and tens of kilometers rover traverses, were necessary and essential for accurate and understandable documentation. Efficient and systematic methods of station characterization and sample collection were driven by severe time and mobility constraints. The efforts and success achieved by the Apollo field geologists on the ground and astronauts on the lunar surface established a precedent and created an effective baseline for all future field work efforts on extraterrestrial surfaces. That baseline includes ground support as well as spacesuit and rover mobility requirements related to EVA durations, allowable rover distances and slopes.

Specifically, how can we apply what we learned then, incorporating technological advances that have occurred since the Apollo era, for use in future field expeditions, human or robotic, to the Moon or elsewhere? Intimately connected to the technology development, or lack of it, is today's greatly changed economic context, including a severely degraded American manufacturing base. Yet this is combined with the availability of far more compact, lower cost, and greatly enhanced performance hardware components and with far more involvement and interest of the private sector in development of a space infrastructure. Even limited to the use of current technology, we could, with aggressive investment over a period of years, establish a sustainable American manufacturing beachhead beyond low Earth orbit, taking advantage of minimal gravity and unlimited access to power and useful materials to produce higher quality specialized glass optics, optical crystals, glass alloys, semiconductor chips, ceramics, superalloys, biofabricated materials including cells and proteins [1].

Thanks to capabilities we developed to support the Apollo program (rendezvous and docking; flyby and/or gravitational capture; orbiting, landing, surface traversing and sampling of solar system bodies; deep space navigation, tracking and communication; life support), we now have far more evolved reconnaissance of solar system bodies from in-depth programmatic surveys and analysis tools and resulting follow-on executed missions, from a far broader constituency. Until recently, the Moon itself had been largely neglected by the NASA planetary science division as being the bailiwick of human exploration. The serendipitous discovery of water on the lunar surface [2, 3] has led to renewed interest in the Moon, reexamination of existing lunar data, implementation of new missions to the Moon [4, 5] (Table 9.1) and strategic shift in viewing the Moon as the new 'high ground'. At the same time, aerospace economic, international, and thus strategic interests have shifted from LEO to the Moon and cislunar space, as the focus of a sustainable space economy and the new 'high ground', allowing increased access to space at lower cost. The next generation exploration will thus involve using the moon and cislunar space as permanent and semi-permanent 'platforms' for many activities requiring full-scale reconnaissance. These include the expansion of the

Table 9.1 Post-Apollo era successful American lunar missions or payloads

Year	Mission	Description
1992	Geotail	Flyby Magnetosphere
1994	WIND	Flyby Magnetosphere
1994	Clementine	Orbiter Surface Composition, Mapping
1998	Lunar Prospector	Orbiter Surface Composition, Mapping
2007	Chandrayaan-1 (M3 payload)	Orbiter surface composition, volatiles Mapping
2009	Deep Impact	Flyby (on way to comet Hartley 2), Volatiles
2009	Lunar Reconnaissance Orbiter	Orbiter Surface Composition, Geology, Mapping
2009	LCROSS	Impact generated subsurface volatiles
2011	GRAIL	High resolution gravity map and interior
2013	LADEE	Atmosphere and Dust environment
2022	CAPSTONE	GPS and long duration orbit pathfinder
2024	IM1, IM2, Firefly Aerospace	Commercial Automated Landers

communication, transportation, industrial, and exploration infrastructure for the Earth Moon system and the development of a proving ground for autonomous operations beyond the Earth Moon system but with the 'fail safe' option of 'real time'. NASA is supporting working groups to develop technologies for use on the lunar surface through the Lunar Surface Innovation Initiative [6].

As far as the strictly lunar exploration is concerned, we have visited wide-ranging targets of interest on the Moon and in the inner solar system and in many cases either performed in situ measurements or collected the first samples; thus, our areas of uncertainty are far more focused. Requirements for study locations are far more specific as well, and typically involve traveling to increasingly inaccessible geological features and landing sites, compensating for the fact that we had targeted the more accessible sites the first time around to minimize risk. Much more comprehensive and detailed basemaps can be created now as the number of missions increases, resulting in greater resolution maps of compositional variations, topography and underlying structure, and greater high resolution photographic coverage. In addition, greatly resource-minimized experiment packages are available. On the other hand, we still don't know the 'lay of the land' for many sites from the perspective of a human on foot or a rover mobility system (on scales of tens of centimeters) except where the astronauts landed, and many of the landing sites of interest now are on far more rugged terrain than explored previously. We certainly have a greater capability for streaming audio, video, instrument feeds, and complex, higher degrees of freedom mechanisms operating in more extreme environments [7], translating into hands free operation for humans and greater sample interface capability for robotic devices and instruments. Extreme miniaturization of electronics, sensor and mechanisms, a trend initiated during the Apollo era, make multi-platform deployments at extremely low altitudes or on the surface possible. Such an approach would necessitate embracing a higher risk and initial investment for supporting infrastructure, in other words truly embracing the 'cubesat paradigm': small, standardized SWAP and interfaces, multi-platform concepts that overcome 'single point failure' risks by the numbers [8, 9].

9.2 In Situ Exploration Tools and Compact Package Networks

The last decade has seen the development of compact yet powerful, multi-functional instruments capable of providing in situ measurements with little to no sample handling (thus reducing power, mass, and risk associated with mechanisms and providing faster, higher resolution, far more comprehensive analysis of samples in situ) [10, 11]. Robust versions of instruments of particular interest ranging from kilograms to ten kilograms integrated with small orbiter, rover or lander platforms, or even handheld by astronauts, are already under development via NASA DALI, CLPS/NPLP, and SIMPLEX programs. These include compact and robust near to IR cameras and imaging spectrometers (e.g., EECam, mini-TES) or multi-spectral micro-imagers to characterize surface composition and properties; neutron spectrometers, magnetometers, sub millimeter sounders, and seismometers (e.g., mini-NS, SEIS) to characterize the subsurface, energetic particle analyzers (mini-ESA, mini-ENA) along with mini tunable laser (mini-TLS) or mass spectrometers (mini-QITMS) to measure in situ gaseous species.

Deployment of a network of compact orbital, lander, and/or rover payload platforms would greatly facilitate the meeting of high priority lunar exploration objectives, ultimately leading to more efficient resource utilization, that requires distributed measurements to be fully realized [10, 11]. Such packages could be used to determine the global distribution, origin, and thus inventories of in situ resources such as water and other volatiles at local-scale resolution. Measurements from these packages would allow monitoring and modeling of the interactions between radiation, charged particles, exosphere species, micrometeorites, surface, and subsurface regolith constituting the lunar environment, and the lunar interior, constraining the Moon's history and origin. Some distributed networks that could be envisioned include: (1) tens of mini-rovers to characterize subsurface water (with IR imagers, mini-neutron spectrometers, and mini-GPRs) to a depth of 1–2 m traveling along traverses at several likely candidates for hundreds of ppm near surface water; (2) mini-landers spaced over hundreds of kilometers that characterize the lunar water cycle (with solar wind analyzer, energetic neutral analyzer or mini-QITMS, and IR imager), and (3) several landers spaced over hundreds of kilometers that characterize the lunar interior (with seismometers, heat flow experiments requiring a small drill for deployment, and magnetometers.

A major challenge for small packages, particularly on the lunar surface, is thermal packaging to protect the payload from the lengthy temperature extremes without the need for active control systems requiring power and thus significantly increasing mass and volume needed for batteries during lunar night. High performance thermal component-based packaging based on passive thermal design that will allow operation on at least limited duty cycle during lunar night are under development [12] and will be discussed further below.

9.3 State-of-the-Art Documentation

None of these capabilities mitigate the need for efficient, systematic documentation and sampling methodology, as used on Apollo. In fact, the need is more critical, because of the limited mass available for sample return, still between 100 and 200 kg with any current architecture, despite the fact the much longer expeditions are planned. This translates into the need for more extensive grading and down-selecting samples of potential interest in situ, through the use of portable rapid analysis tools as described above, as an additional step in the sampling and site characterization process developed for the Apollo J missions. For robotic missions, the application of the most productive field methodology would be through the use of telerobotics, an expensive and technologically challenging option that has been discussed but not seriously considered to date. Alternatively, the development of rovers with greater (faster, more autonomous, more rugged terrain gaits) mobility, such as the shape–shifting 3D reconfigurable tetrahedral rovers proposed over a decade ago [13] would allow access to a larger variety of targets of interest in a given time frame. This would greatly facilitate field work by creating opportunities for the human interaction and participation so effective in sample selection and site characterization at a more human intellectual pace.

9.4 Power and Thermal Challenges

The lunar surface has extreme (1 month) diurnal cycle induced temperature variations [14] with extended (2 week) hot and cold periods at low latitudes, where temperature vary between 400 and 100 K. At the lunar poles, days and nights are 6 months long, just as they are at Earth poles, with the sun always at relatively low sun angles and temperatures varying between 175 and 50 K. Currently, NASA is funding work to develop more efficient, long term solar power generation at the South Pole, planned site of a long-term human occupied station [15]. Solar cell arrays, called VSAT for Vertical Solar Array Technology) would be resistant to abrasive dust, stable on sloped terrain, and on tall masts to maximize power generation when the sun is particularly near the horizon.

Solar power (during periods of illumination) combined with radioisotope power supplies (RPS) and thermal units (RTU) provided power and heating during lunar night for earlier lunar surface missions, but the costs are high and ready availability of radioisotopes in the future is uncertain. Radioisotope based power systems are extremely inefficient, necessitating greater mass and volume and thus higher costs. Greatly resource-minimized experiment packages could be available, if the development of components (e.g., batteries, electronics) that can operate at ultra-low power and withstand ultra cold temperature operation continues at a reasonable pace [16] and the next generation high performance thermal components (switches, radiator/reflectors, heat pipe embedded structures) are utilized [12]. Over a decade ago now, work was being done on CMOS (complementary metal oxide semiconductor) components to support microminiaturized System on Chip applications running at lower logic voltages (0.5 V) instead of the standard 5 V, resulting in 100:1 reduction in power requirements. 0.35 um CMOS Ultra-Low Power Radiation Tolerant (CULPRit) technology was successfully tested and validated on the ST5 mission. Achieved by aggressively scaling the transistor threshold voltage and supply voltages so that the circuit performance, in principle, remains constant the chip was observed to be fully functional in a Reed-Solomon encoder down to 17.5 K. This was followed by development of a 0.18 um CULPRiT technology applied to a General Dynamics 32-bit coldfire microprocessor reengineered to run at 0.5 rather than 3.3 V [17, 18]. The demonstrated CULPRit ability to operate below 25 K significantly reduces the thermal and thus power required for operating with MilSpec electronics which have a minimum operational temperature of 223 K.

While challenging, the radiation environment on the Moon is far less extreme than the conditions that will be experienced by Europa Clipper mission. Low power, efficient, radiation hard MOSFETS (Field Effect Transistors, a very efficient subclass of transistors) developed by Infineon, extensively used in Earth orbiting satellites that require enhanced strategic protection from hostile disruption from weapons systems, are also being used for the Europa Clipper mission. Further advances are currently being made through the use of Silicon Carbide rather than Silicon based components, and/or the use of CMOS Field Effect Transistor designs such as Silicon on Insulator (SOI) or FinFET (Fish Fin shaped). These architectures greatly limit pathways for radiation-induced charge disruption [19]. Remaining major limitations are the limited discharged rates and narrow operational temperature range of conventional batteries.

The near surface mobility necessary for next generation reconnaissance would benefit greatly from the development of ultracapacitors as replacements for conventional batteries. Like RadioThermal Generators and unlike conventional batteries, Ultracapacitors can operate over a wide range of thermal conditions. Ultracapacitors, with the incorporation of Voltage Dependent Resistors (VDRs) [20] could provide both peak power at high discharge rates for energetic maneuvers (e.g., climbing slopes) and steady power over 'flat surfaces' at lower discharge rate. NASA MSFC have been working on a possible solution for mobility systems: a 'hybrid' ultracapacitor concept which combines the properties of high capacitance battery-like discharge [21, 22]. These devices rapidly discharge down to a certain voltage, and then discharge more slowly below that voltage.

In addition, the advancement of the cubesat paradigm beyond LEO into the more extreme environments of cislunar space over the last decade has led to the development of a range of high performance components for thermal control. These include [12]:

- Thermal Switches: high efficiency Reverse-Operation Differential Thermal Expansion Switches (ROD-TSW) that passively transition from 2 W/K at 300 K to 0.0015 W/K at 260 K (ON/OFF ratio = 1333) (with mini-loop heat pipe (LHP) added on FSS to increase efficiency)
- Spacerless MLI: multi-layer insulation (MLI) blanket without spacers utilizing cryo-system construction methods (Vectron tension cables to hold in place) and 30–40 layers of double aluminized Mylar with dacron separators to attain a very low effective emissivity
- Isolating Mounts: high performance Ultem 1000 isolating mounts to attain a very low conductance from the enclosure wall to the lunar lander deck.
- Mechanical support: Heat pipe embedded structures for cooling spacecraft interiors
- Unconventional radiator design: Incorporation of radiator/reflectors for prevention of heat transference from the sun and lunar surface simultaneously, variable thickness radiators, and deployable radiators

Extreme cold temperatures affect the operation of mechanisms as well as electronics. Robotic rovers particularly depend on these mechanisms to survive and operate in during lunar night or in permanently shadowed regions to explore and characterize lunar volatile cold traps. JPL has developed a solution: Bulk Metallic Glass Gears capable of operation under such conditions [23]. Bulk metals are completely melted to randomize their initially organized crystalline structure and form an amorphous material behaving more like glass capable of being injected molded. BMGG has much higher strength and wear resistance and is much less brittle than ordinary metal, which is what is needed for gears to operate without lubricant below 200 °C.

9.5 Next Generation Rovers

Exploration of the Moon with uncrewed vehicles (Table 9.2) did not cease when the Apollo program was cancelled, and in fact has increased through the years, as unanticipated discoveries have continued to make the Moon more intriguing from scientific, economic, and strategic standpoints. JAXA LEV-2 is actually ball-shaped and can lengthen and wiggle from side to side.

A return to the Moon, on a more permanent basis, will require more comprehensive study and ongoing monitoring of the environment, identification and utilization of in situ resources, reconnaissance and situational awareness to support uncrewed or crewed 24/7 operation. Such a goal will require maintenance of a communication and transportation infrastructure.

A variety of conventional wheeled robotic rovers have been proposed for lunar exploration, varying in size and in the degree to which they work synergistically with human astronauts. JPL, with its extensive experience in the development of robotic rovers, has flown a number of long-range lunar rover concepts (Table 9.3). Most recently, JPL has developed Cooperative Autonomous Distributed Robotic Exploration, CADRE, a distributed network of mini-rovers (Fig. 9.1) with the plan of deploying them to map an aera in Reiner Gamma using an Intuitive Machines lander [24]. Motiv developed the lander deployment system, motor controllers, exoskeleton, and thermal management system. Ongoing monitoring of the motor frequency allows ground control to determine when the a particular mini-rover needs maintenance. Astrobotic has also been testing CubeRover, a cubesat-scale rover [25].

Meanwhile, NASA has funded the development of concepts for unpressurized rovers [26], potentially capable of operation with or without humans onboard, to support surface transportation for exploration of the lunar

Table 9.2 Uncrewed lunar rovers

Date	Rover	
1970	USSR Lunokhod 1	Landed by Luna 17, first extraterrestrial rover, Mare Imbrium
1973	USSR Lunokhod 2	Landed by Luna 21, Mare Serenitatis
2013	CSNA Yutu	Chang'e 3 Mission, Mare Imbrium
2019	CSNA Yutu-2	Chang'e 4 Mission, South Pole Aitken Basin (farside)
2023	ISRO Pragyan	Chandrayaan-3 Mission, Nearside South Polar Region
2024	JAXA LEV-1	SLIM lander, hopper, eastern equatorial nearside
2024	JAXA LEV-2	SLIM lander, reconfigurable ball, eastern equatorial nearside
2024	CSNA Jinchan	Chang'e 6 Mission, South Pole Aitken Basin (farside)

Table 9.3 JPL robotic rovers

Program	Year	Mission description
Mars Pathfinder Sojourner	1997	Cameras, Demo Faster, Better, Cheaper, days
MER-A Spirit	2004	Camera+mineralogy payload, years
MER-B Opportunity	2004	Camera+mineralogy payload, years
MSL Curiosity	2011	Larger, Comprehensive payload, ongoing
Mars 2020 Perseverance	2020	Larger, Comprehensive payload, ongoing

Fig. 9.1 JPL CADRE mini-rover with Mars Perseverance rover engineering model in background. (Credit: NASA JPL-CalTech)

surface (Fig. 9.2). Specification details are not yet available, but the idea was to incorporate the most efficient solar panels, lightweight rechargeable (between tens of kilometer trips) battery packs, larger wheelbase and seating higher above ground for greater stability and safety, and mobility

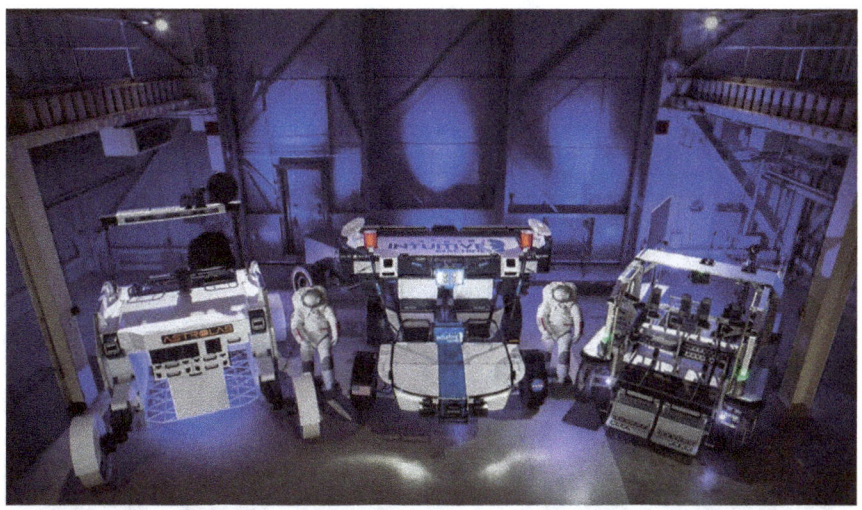

Fig. 9.2 NASA Concepts for unpressurized rovers from left to right Lunar Outpost Eagle, Moon RACER, Venturi Astrolab FLEX. (Credit: NASA/Bill Stafford)

Fig. 9.3 NASA Concepts for pressurized rover: JAXA Cruiser. (Credit: NASA and JAXA)

characteristics comparable to the Apollo Lunar Rover. Available images indicate they are larger in size (and mass) than the Apollo rover, with RACER being the most compact. In addition, NASA has asked JAXA to develop the concept of a pressurized rover concept called Cruiser [27] (Fig. 9.3). These rover concepts are described in Table 9.4.

What about greatly enhancing mobility capabilities, to include three-dimensional operation to allow access to highly variable and rugged terrains?

Table 9.4 NASA Artemis rover concepts

Developer	Rover
Intuitive Machines	Moon RACER Reusable Autonomous Crewed Exploration Rover
Lunar Outpost	Eagle
Venturi Astrolab	FLEX Flexible Logistics Operation
JAXA	Cruiser (Pressurized)

Fig. 9.4 Intuitive Machines proposed Micro-lander 'Nova-C' capable of multiple hops on the lunar surface as described in the text [29]. (Credit: Intuitive Machines)

Work has been done to expand the concepts for near surface mobility beyond conventional wheeled rover locomotion, as described below. Unconventional systems, more complex than wheeled axle systems operating in a two-dimensional surface framework, would require greater autonomy and a longer 'training' or learning curve. The concepts described below are mini-rover size, capable of carrying kg-scale payloads, which could certainly include high resolution camera and tiny environmental (temperature) and surface property (electromechanical) sensors, but potentially scalable to increase payload capacity.

Locally operated hoppers or drones have been a popular approach for increased mobility [28] (Figs. 9.4, 9.5, 9.6, and 9.7). Because Mars has an atmosphere, albeit thin, it is possible to use a very large blade helicopter, rather than hoppers, to travel across the surface for long or short distances there, but that is not so on the atmosphere less moon. Most frequently, small chemical propulsion driven mini-landers that are capable of making several

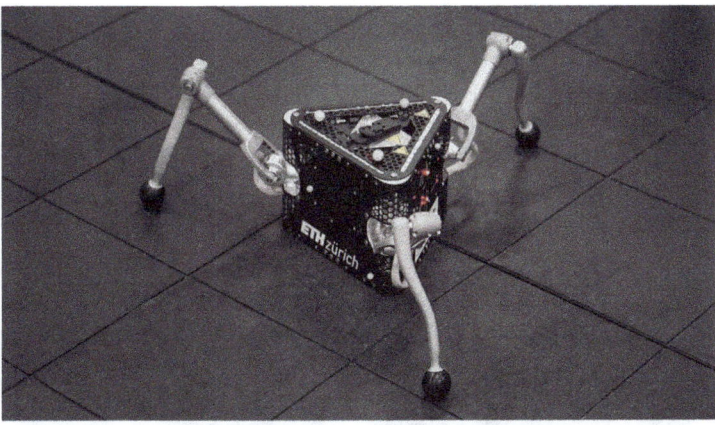

Fig. 9.5 Model of ETH Zurich battery powered gimballed-leg Tumbler capable of multiple jumps as descried in the text [31]. (Credit: Evans, IOT World Today)

Fig. 9.6 MIT proposed Hovercraft with Ion Propulsion Drive for exploring the lunar surface as described in the text [30]. (Credit: Catwell, Element 14)

small 'hops' of varying lengths have been proposed for the Moon. The possibility for refueling drones (via a lander or more permanent base) has been discussed. One example is Intuitive Machine's proposed MicroNova, a chemical micropropulsion drone to be mounted on their Nova-C lander capable of carrying a 1 kg payload, including a high-resolution camera, up to several kilometers [29]. MIT has prototyped a hovercraft with ion propulsion

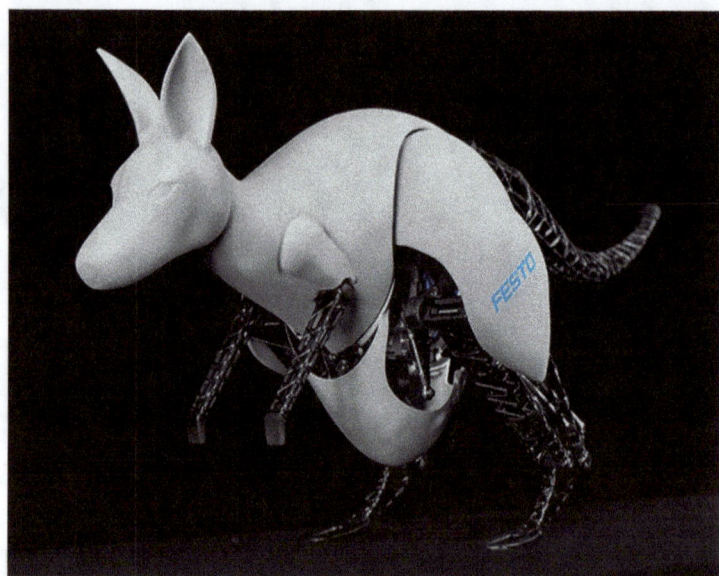

Fig. 9.7 Model of Festo hopping kangarobot as described in the text [32]. (Credit: Del Guercio, Science Today)

thrusters [30]. ETH Zurich developed a lightweight battery powered device with three gimballed, jointed legs suitable for low gravity environments. In motion it behaves similarly to a cricket, pushing off, tumbling, and then, through use of orienting gyroscopes, landing the payload 'upright' [31]. Not surprisingly perhaps, as kangaroos have the most efficient long-distance travel gait known [32], Elias Knubben at Festo, a 'smart' pneumatics company, has developed a kangaroobot. Just as a kangaroo does, the device has a very stable hopping mode capable of adjusting for variations and yet maintaining in a stable upright orientation [33, 34]]. The hop is accomplished by pretensioning elastic spring elements with the aid of pneumatics.

A yet more revolutionary approach, allowing rapid three dimensional reconfiguration in shape and size, is the Tetrahedral Explorer Technology (TET) [35–40] (Fig. 9.8) based on the tetrahedrally connected nodes and struts as a 'building block' and advanced autonomic level autonomy. This highly autonomous rover is designed to be capable of learning gaits suitable for any terrain it will encounter as described below. Conformable tetrahedra are the simplest space-filling form the way triangles are simplest plane-filling facets. TETs are shape shifting mobile platforms with reversibly deployable struts forming edges and connecting via nodes at apices. Conformable tetrahedra are the simplest space-filling forms the way triangles are simplest plane-filling facets. The 12Tet (figure) consists of 26 struts and 9 nodes forming 12

Fig. 9.8 NASA/GSFC Concept for 3D reconfigurable tetrahedral rover as described in text. (Credit: NASA/AMA Productions)

interlinked tetrahedra. These undifferentiated (with no permanent appendages such as wheels) interlinked forms have the degrees of freedom necessary to develop a variety of gaits from simple rolling to crawling or reaching, and, at their most efficient, merely extending the center of mass forward around nodes on the ground and tipping in a spiraling gesture. This is inspired by pseudopodia formation in amoeba, a most efficient 3D locomotion mechanism [35]. Control is a key challenge in realizing a rover with a highly addressable structure that can operate in highly irregular terrain. The 12 TET rover is designed to become a moving part of the terrain, its multi–nodal ranging system providing volumetric information about its surroundings. Gaits suitable to rugged terrains and metrics to measure performance are being developed and tested. We initiated the development of TET operational scenarios by considering mobility requirements as a function of terrain and by developing test gaits, such as the 'amoeboid gait'. The most recent prototypes were teleoperated, requiring the design of macros for coordinated movements, definition of actuator commands and incorporation of sensor telemetry parameters as feedback for the actuation process via a wireless communication scheme through a user interface.

9.6 Advanced Autonomy for Locomotion

Tetrahedral systems [35–39] are designed to be capable of autonomic autonomy, deploying and stowing struts in a 'gaited' manner learned in advance to be appropriate for a terrane or application and then modified within a dynamic yet stable envelope to adapt for actual conditions each step of the way. Feedback is provided by touch and motion sensors on each node combined with a bilevel navigation system as described below. The highly addressable structure must operate in highly irregular terrain, where locomotion requires an intimate blending of dynamics and statics. Most means of locomotion try to "finesse" the situation by somehow glossing over the complexity of the terrain: typically, rovers have featured wheels or legs that are larger than the terrain scale sizes, for locomotion that is slow to allow expert computer systems time to figure out where to go next. The tetrahedral rover (TET) becomes a moving part of the terrain, its vision system providing volumetric information about its surroundings. With information about the geometry of its environment as well as information about its own geometry, the TET places itself within and moves through its environment.

Tetrahedral devices require a robust vision/navigation system with minimum power and bandwidth requirements which has yet to be developed. The system would be required to provide feedback rapidly while the vehicle is in motion, locate small obstacles within meters of its immediate path, such as boulders, as well as remote hazards tens of meters away in the direction of motion such as cliffs. The candidate 3-D Vision system that best meets our requirements is already under development for a wide range of deep space, orbital, and surface exploration applications. It is laser-based 'scannerless' range imaging system consisting of a laser diode emitter array, low power high-resolution time-of-flight ranging electronics, and a mega-channel fiber-optic based receiver. All of these components are extremely efficient, compact, and robust, and would combine to make a highly reliable 3-D imaging system. We envision this as a bilevel imaging system combining short-range, high-resolution imaging to support local maneuvering and immediate hazard detection; and longer range imager for trajectory planning and large-scale hazard avoidance.

Ultimately, autonomy will be achieved through a neural basis function (NBF) software architecture used reactively to define, control, and organize the network of actuators responsible for TET motion. The patented [41] ANTS (autonomous NanoTechnology Swarm) AI. Neural Basis Function (NBF) NBFs are composed of multiple control systems within an Evolvable Neural Interface (ENI) which acts as an active communication medium between the control elements. Separate behaviors of the TET are typically instantiated as separate NBFs: to add new behaviors the aim is to simply link the ENI of the new behavior into the system and then allow the ENIs to adapt the old and new components to each other. Genetic algorithms dynamically generate gaits (a series of actuator deployments) in response to sensory input in a simulated environment appropriate for a given landscape (such as the Moon). The behavior of the rover would be dynamically controlled by stability algorithms to maintain the system within operational limits. ST8 Autonomous Rendezvous and Capture algorithms are being used to develop the synthetic neural system.

Target selection can, and is now currently, occurring via human interface (telepresence). The next step would be to develop a virtual reality model and high-level command language for user wireless communication interface. This autonomically smart system, which allows the rover to travel from target to target without external direction, can ultimately be linked to higher level heuristic (decision making for target selection) intelligence system through an evolvable interface, a system with bilevel intelligence, which would mean carrying high level intelligence capability onboard for entirely robotic operation. The high-level components generally rely on a more symbolic approach to control and may involve planning and schedule and other heuristic control. The NBFs for the TET are built on what we have learned by applying the NBF architecture to a control system for autonomous rendezvous and capture of a chaotically tumbling target, a problem inspired by the Hubble Space Telescope Rescue mission.

The great advantage of autonomous locomotion as described above is the potential for true 'discovery' mode of operation. Is the Apollo methodology adaptable enough to provide 'lessons learned' for a 'target of opportunity' approach? Actually, the Apollo methodology was quite flexible. Initial path planning and preliminary geological mapping of the landing area was accomplished with the highest resolution data available and the crew trained using a simulated landing site approach, to protect the safety of the crew as well as to maximize crew productivity during landing site reconnaissance. However, the crew sampled, documented, and made measurements 'at will' at a site of interest, and could modify the route, eliminating or adding stations as well as increasing or decreasing the amount of time spent at designated targets. This approach took advantage of the type of training they received and at which humans excel, intuitive grasp of a site in the context of the surrounding landscape in situ. Could this human capability be combined with capability for autonomous movement in rovers without humans in the loop for targets selected by a human crew in real time? What can be seen from boots or wheels on the ground is not the same as what can be seen from overhead. In this case, development of Terrain Relative Navigation capability, discussed below, would be crucial [42].

Although plan view is useful to keep track of location and coverage of an entire area of interest, the location and accessibility of actual samples and terrains is made possible by the small–scale 3D views visible on the ground. A true 'target of opportunity' or 'discovery' approach with less stringent requirements for advance knowledge or planning would be made possible when risk is mitigated by the alleviation of severe resource restrictions, by for example, having in the field a fleet of self–sustaining, autonomous, reconfigurable mobile devices with dynamic architecture. Field work on the Moon accomplished during Apollo established a baseline for human performance, surface mobility (with humans in the loop), and useful site characterization (with or without humans) on rocky bodies with reduced atmospheres regardless of architectural style, exploration scenario, or target.

9.7 Advanced Autonomy for Dynamic Structures

The same Autonomous Reconfigurable Technology (ART) principles used in the tetrahedral rover could be used as the basis of dynamic structures for a mobile architecture [35–39, 41–43]. The goals, as in the rover, is to create autonomic, reusable, reconfigurable, highly redundant, efficient, and thus cost-effective systems which could create the multi-functional infrastructure. ART architecture is realizable in stages at any level of electromechanical system technology (macro (EMS) to mini (MEMS) to nano (NEMS) and promises to revolutionize space structure design by epitomizing portable 'form follows function' at every level (Fig. 9.9). The tetrahedral rover, the basic building unit in this architecture is a tetrahedron, the most efficient space-filling form, consisting of nodes interconnected with struts that can be reversibly and/or partially deployed or stowed to allow the tetrahedron to change its size and shape on command in seconds. Struts are formed into tetrahedral units. Units can be interlinked in one (linear), two (planar), or three (space-filling) dimensions to create conformable objects. As more tetrahedra are interconnected, the degrees of freedom are increased and motions can evolve from simple to complex, from stepped to continuous.

Early work on ART prototypes (Tetrahedral Walkers) provided a basis for understanding performance and conceptual models for hardware and software requirements which could support the architecture proposed here [35–39, 41, 42, 44]. Simple prototypes for mobile units were constructed using cable and pulley and then screw drive deployment mechanisms at the electromechanical systems (EMS) level. Hydraulic or pneumatic mechanisms could be used to control telescoping struts, but they would require more mass and volume. As Micro-EMS (MEMS ART) or nano-EMS (Super Miniaturized

Fig. 9.9 Concept of reconfigurability and multifunctionality of ANTS-based robotics. Top: Animation of EMS (macro) level tetrahedral rover (left to right) climbing, bridging, flattening, and clambering over obstacles in natural terrain. Bottom: Animation of MEMS level ANTS-based dynamic infrastructure (left to right) landing, maneuvering around obstacles, communication. (Credit: NASA/RASC project)

ART or SMART) become available, within the next decades, such components could be incorporated to minimize mass and power requirements. The 3D network of actuators and structural elements is composed of nodes that are addressable as are pixels in an LCD screen.

As described above, ART structures are capable of mobility in extremely rugged environments and don't require the flattened surfaces preferred by wheels. However, 3D ART structures could form bridges or tunnels in particularly rugged terrain (e.g., large ravines) to make that terrain efficiently accessible or usable for other purposes, such as resource extraction operations (Fig. 9.9). Components could be deployed for a particular function, stowed and transported to a new location, then transform into another form as required for a new function. The ultimate goal would be to create structures capable of routine self-servicing, including recharging, retrieving, replacing, recycling components as required in a cost/effective manner. The particularly efficient amoeboid gait would be used in 3D, as the amoeba itself uses it, to minimize energy consumption.

9.8 Communication and Navigation Challenges

Recognizing that the demand far more robust communication, navigation and tracking capability will be required to support an on-going presence in cislunar space, NASA plans to greatly expand the Deep Space Network (DSN), originally built for the Apollo program, and now known as the Space Communications and Navigation Network (SCAN). Enhancements for cislunar space, discussed below, include Lunanet, a lunar global scale delay tolerant, Monitored beacons, a lunar GPS, and multiplexed communication.

Lunanet: The cislunar neighborhood network [45] would offer services in cislunar space comparable to those now offered on Earth.

- Rather than be reliant on prescheduled links between ground and space assets, the expanded network will allow connectivity for communication, navigation, and data services on demand via direct or relay links, utilizing a technology known as Delay/Disruption Tolerant Networking (DTN) being tested now. DTN enables nodes along a communication relay path to store data until a signal disruption disappears.
- Lunanet architecture for lunar navigation would allow high precision onboard orbit or surface position determination and guidance system operations independent of data processing on the Earth for crewed or robotic missions.
- Lunanet will provide alerts and enhance situational awareness for any assets operating in cislunar space, and provide updates of potentially threatening conditions or hazards, such as enhanced solar activity, in real time.
- Lunanet would link nodes providing ongoing measurements on the moon, in and beyond cislunar space and thus allow ongoing monitoring and a baseline for scientific study of the lunar global and regional environment as well as celestial objects in and beyond the solar system

Beacon Monitoring: The DSN is in the process of developing and testing the use of low cost, small antennas in the network to monitor the status of cislunar orbital or ground assets using four tones [46] from onboard beacons. The tone frequencies would be monitored constantly, but the assets themselves would initiate beacon communication, using one of four tones indicating the level of urgency for contact, including 'OK' (track on planned schedule), track as soon as convenient, track at a certain time, or track as soon as possible (emergency).

MPSA: The DSN has also developed a capability called Multiple Spacecraft per Aperture (MSPA) to allow signals from several spacecraft (each operating on different frequencies) within the antenna beam to be received (downlinked) simultaneously [47]. Transmitting (uplink) can still with only one spacecraft at a time. Also possible is the combining of several antennas at a DSN station (arraying) to create a larger virtual antenna and thus increase the downlink rate.

Cislunar GPS: The equivalent of a lunar Global Positioning System, ultimately essential to support operations with a network of assets around the Moon, is being tested by the pathfinder mission CAPSTONE (Cislunar Autonomous Positioning System Technology Operations and Navigation Experiment) [48]. The 12 U cubesat is in a stable, low fuel consumption Near Rectilinear Halo Orbit, comparable to the orbit planned for a future cislunar space station (Gateway). CAPSTONE is validating the capability to operate its Cislunar Autonomous Positioning System (CAPS) in a complex lunar orbital regime (to support a range of orbital, surface, and LaGrange point assets) on a long-term basis. CAPSTONE achieved the planned orbit and, in collaboration with NASA GSFC, is demonstrating this capability by using the Lunar Reconnaissance Orbiter.

Plan for Artemis Program: Currently, NASA plans to equip the lander and spacesuits with a Nokia 4G level connectivity network to support astronaut surface communication for Artemis 3 [7].

Terrain Relative Navigation. What enhanced navigation capabilities should we focus on for next generation rovers to expedite exploration of the lunar surface and prepare for similar activity on Mars? Uncrewed rover mobility would benefit greatly from improved terrain relative navigation, increasing distance covered (and surface characterization) per unit time. The current maximum distance traveled is 150 m/day for Mars robotic rovers. What would be required to increase this by one to two orders of magnitude, kilometers to tens of kilometers per day, more like the performance of the Apollo crewed rovers?

- more rapid traverses over longer distances not only in general, but between human-in-the-loop interventions.
- traversing and sampling more diverse terrains within a single mission, with particular interest in visiting more rugged mid-latitude and polar terrains.
- more autonomous driving, not only with less reliance on 'human-in-the-loop', but with improved ability to identify features of interest 'on the fly' for drive-by science (analogous to 'target of opportunity').

- Maintaining the capability to deal with risks such as terrain (rock size and frequency, regolith mechanical properties) variations, and rover structural hazards

JPL's Machine-Learning Analytics For Rover Systems (MAARS) is the current state-of-the-art for rover capability, applying machine learning for navigation, risk avoidance, and scene interpretation using a vision-based system that combines stereo imaging with NAVCAMs [49]. MAARS utilizes Soil Property and Object Classification (SPOC), a deep-learning based instance segmentation algorithm, an approach typically used for classification of each segment of the scene according to type in a vision-based system [49]. In this case, segments are identified as objects of scientific interest, or hazards (e.g., large rocks). However, the system lacks the capability to identify nongeometric hazards, such as small embedded rocks which could damage wheels, or fine sand which would induce wheel slippage. Currently JPL rovers use in-situ mobility data to generate and empirical wheel slip prediction model. The grayscale NAVCAMs are used to distinguish and detect rocks from their surroundings, stereo imaging is used to estimate rock size, and, by rotating the scene until the surface slope is zero, the rock height. Recognizing the need to maximize science return over greater distances with minimal human-in-the-loop interactions in bandwidth limited environment, JPL proposed to utilize Scientific Captioning of Terrain Images (SCOTI), a tool based on the "Show, Attend, and Tell" model [50], an attention-based model for image caption generation. Using this approach, the rover would be trained, using standard backpropagation techniques with stochastic gradient descent utilizing adaptive learning rate algorithms, on the natural language inputs of geological experts. Utilizing the SCOTI tool and these inputs, including relevance of representative Mars surface images, the rover would have the capability to generate captions in natural language for images obtained along the route as well as taking into account distance (derived stereo data) when determining differences or similarities between scenes. This information would be downloaded and packaged into a searchable primary interface, which could be queried and updated with additional criteria by scientific users, used as an investigation ongoing planning tool, as well as the basis for automated rover downselect of images of highest scientific relevance. More recent work at JPL on uncrewed lunar rover navigation has emphasized matching craters detected on the ground with known craters mapped from orbit. Simulations of this approach have indicated the ability to locate a rover within 5 m [51].

Other areas that show promise are work done on autonomous cars that switch from ground level operator to overhead sensor view, and other approaches that convert the area in front of the sensor (road image) at a known DEM location to map view (change angle). These include a 'birdseye view' algorithm in mathworks [52], and streetview image generation [53].

9.9 Dust Mitigation Challenges

Successful exploration of most planetary surfaces, with their impact-generated dusty regoliths, will depend on the capabilities to keep surfaces free of the performance-compromising dust [54, 55]. Once in contact with surfaces, whether set in motion by natural or mechanical means, lunar regolith fines, or dust, and by implication other the regoliths of small bodies, will coat surfaces, clog mechanisms, make movement progressively more difficult, and be difficult to remove mechanically due to the essential regolith characteristics described in Table 9.5.

During Apollo landings, extensive locally-induced stirring of the regolith caused dust to be suspended long enough to come into contact with conducting surfaces. Dust behaved like abrasive Velcro: it adhered to everything and attempts to remove it by simply brushing did not remove fines (<10 μ) and resulted in severe abrasion and compromised seals. Lunar fines, because of their electrostatic charging, were more difficult to collect in sample bags than larger size particles. Because of this difficulty, details on the most problematic

Table 9.5 Essential regolith characteristics

Characteristics	Description	Analogous simulant?
Volatile component	Water may be lattice-bound, adsorbed, discrete, difficult to 'wet'	Off Planet (OP) 'vapor deposited'
Space weathered 'skin'	Effect of solar wind and micrometeorite (thermal) bombardment	Mixed in nanophase iron with some versions JSC, but not 'skin'
Grain morphology	Impact bombardment results in highly fractured, melty, glassy shards, 3D complex, elongated, high surface/volume ratio	NULHT and OP closest. Hard to simulate hitting/smashing rather than grinding/crushing
Size particle distribution	Impactors large to small scales produces large range ≪1 micron to tens of microns	OP closest. Hard to get fines 'tail' and could be major source of characteristics
Agglutinate portion	Impacts produce amorphous, melted fragments	OP
composition	Igneous rock suites: basalt/mar volcanism, anorthosite/gabbro, Mg Suite/Fra Mauro Basalt/KREEP	all

lunar dust particles are relatively sparse [56, 57]. These issues must be resolved for future missions [58]. Movement of dust through natural means (terminator approach) or mechanical contact observed on lunar surface [59–62]. What we do know is that fields, charged particles, and dust particle interactions on Moon complex, dependent on highly variable environmental conditions and particle properties (size, composition, shape, magnetic or electrical parameters).

- Lunar fines have low electrical conductivity and dielectric loss, and tend to remain electrostatically charged [57].
- Charged dust grains are repelled from like-charged surface or attracted to oppositely charged surface.
- Greater illumination and temperature increase surface potential.
- Electrostatic charging occurs via interaction of solar UV light with surface causing photoemission of electrons and interaction of the local plasma environment [63].
- Surface charging on the dayside driven by photoelectron currents, resulting in electron depletion and positive charging of surface; on nightside, plasma electron currents result in electron accumulation and negative charging of surface [64].

Since the Apollo program, a wide range of active and passive approaches for removing, collecting, or controlling lunar or asteroid regolith have been proposed, including active methods such as brushing, pneumatic 'vacuuming', pressurized gas, and electrostatic approaches, or passive methods involving mechanical barriers, materials with surface structures that prevent lodging of dust particles, filters, or seals [65]. The first approach used during the Apollo program, brushing, was not only ineffective in removing fines but was observed to abrade the pressure suite fabric as mentioned above. Lotus surfaces [65], which have lightweight nanotextured coatings, are not only superhydrophobic, like lotus leaves, but prevent particles from sticking. They are also stable in vacuum over a range of environmental conditions, with low outgassing. The coating can be easily applied to almost any surface. The fractal sizing of nanoparticles prevents contaminants from having contact with spaces in between the tops of particles [66].

Seals being considered include spring-loaded Teflon and Aeroflex shaft seals used on the Mars Exploration Rovers, though their durability in the more extreme lunar environment is debatable. The magnetic coupling seal used in terrestrial space environment may attract magnetically susceptible particles in the lunar environment.

Pressurized gas has already been used to blow fine particles onto collecting surfaces. PlanetVac, developed by Honeybee Robotics, is designed to do this with a down-thrusting powerful air jet and large area to collect displaced regolith above it [67]. Missions such as OSIRIS REx have already used this approach on a smaller scale, demonstrating that this approach is not an efficient mechanism for dust collection/removal. The gas itself is a consumable and a concern is abrasion of surfaces by gas-driven particles.

Lunar dust is dielectric with low electrical conductivity, complex surface morphology, enhanced magnetic susceptibility, and tends to remain electrostatically charged. Electrostatically based approaches to dust removal have been considered before and shown some success in moving dust, but the sticking mechanism must by strong enough to overcome van der Waals force and gravity. Simulations of dust behavior in a vacuum have already demonstrated spontaneous dust aggregation in an environment with weak charging, and disaggregation in the presence of strong net positive or negative charges [68].

Several groups have proposed active electrostatic approaches [68–70]. The successful dust removal strategy will have to deal with dust dynamics resulting from interaction between Van der Waals and Coulombic forces (EMP dust). Whereas Van der Waal's force is correlated with the inverse square of the distance, the Coulombic force depends on the surface charge density regardless of distance. The larger surface area and reentrant structure of impact-generated dust grains make them more susceptible to Coulombic forces.

The successful electrostatically based concept would involve controlling the charge transported to a dust covered surface to (1) induce dust repulsion and thus flow from that surface to (2) surfaces of lower potential for removal. Rovers could be fitted with devices that could harness the removal of dust for sampling as part of the extended exploration process on Mercury, Mars, asteroids, outer solar system satellites or the Moon.

The SPARCLED (Space Plasma Alleviation of Regolith Concentrations in Lunar Environment by Discharge) concept [54, 68] is a compact, low power (<5 W), low mass (<5 kg) electrostatically-based device designed and demonstrated to control (remove and collect) dust flow in a simulated lunar environment (Fig. 9.10). SPARCLED, based on heritage from ionic sweepers for active spacecraft potential control, uses an approach analogous to charged beam sweepers controlling the flow of potential across spacecraft surfaces in a highly charged environment. The SPARCLED experiment tested the interaction between an electron gun and dust particles (<20 μ JSC-1AF simulant) in the presence of a weak electric field in an environmental chamber with intermediate vacuum (10^{-5} to 10^{-6} atmospheres) simulating an airlock. The electron-gun generated electron beam (milliamp range) was focused using

Fig. 9.10 NASA/GSFC Concept of SPARCLED handheld electrostatics-based dust removal tool as described in text [54]. (Modified from Clark et al., Fig. 8, 2010, AIP Conference Proceedings)

a − 900 V VDC grid and a + 1000 V pin probe 2 mm above the non-conducting dust-covered surface in the presence of a moderate electric field (500–1000 V). The introduction of a weak electric field (500-1000 V) initiated a cascade of electrons in the beam by ionizing the surrounding low-density gas, greatly increasing the effective flux and luminosity of the beam. The negative charge/mass ratio of the initially neutral dust grains rapidly increased, causing sufficient electrostatic repulsion to accelerate grains rapidly away from the negatively charged surface and implant them on the surrounding relatively positively charged chamber walls. This approach overcame van der Waals forces, and dust removal from collection surface should be possible when its potential is considerably increased. These results were preliminary and further experiments must be performed to determine optimal potential for collection surface, and the range of flux, field strength, and gas density over which this mechanism will operate, by varying electron gun voltage, electric field strength, wall potential, and vacuum settings over a range of dust grain sizes and compositions.

The relatively small amount of Lunar soil samples collected are being carefully preserved and not being used for simulants. In order to test the viability of equipment for longer occupations of the lunar surface, we must create simulants. In fact, so far, the simulants produced have generated from terrestrial igneous rocks with compositions within the range of lunar mare and highland rocks are used. More importantly, can we create simulants that represent the morphological and size distributions of lunar soils which result in their physical and electrostatic properties? Table 9.5 summarizes properties of lunar regolith and comparisons to most analogous simulants. Regolith simulants to date include the ubiquitous versions of JSC-1, EAC-1, Zybek and USGS versions of NULHT, Off Planets OPRL and OPRH non-agglutinate and agglutinate series, Exolith Lab LHS and LMS series of highland and mare soils and agglutinates. In comparison tests, the soils which have physical appearance and behave most like lunar regolith in terms of packing, cohesivity, and size distribution is the NULHT. Unfortunately, very little of this simulant was made. Off Planet simulants are also close to lunar regolith soils in appearance, but were not comparison tested. The biggest constraint to moving ahead to make more of this analogous material is cost.

9.10 State of the Art Astronaut Gear

One critical area where lessons learned from the Apollo program should be applied is astronaut gear, to make EVA time as supportive of astronaut productivity as possible. Work continues to be done in two areas in particular: the pressure suit and enhanced capability gloves. Axiom Space was selected by NASA to develop a prototype for the Artemis era suit appropriately called the AxEMU, the Axiom Exploration Extravehicular Mobility Unit, which would provide much more flexibility at all the joints. AxEMU has an outer layer designed to repel dust, connection to a more integrated sensor net, and a more comfortable fit for men and women [71–73].

In partnership with GM, NASA has supported development of a force magnified robotic upper arm. This development has already been applied in industrial workplaces to increase efficiencies and reduce injuries and could be employed in robotic arms on the moon or even within the spacesuits themselves. The glove enhances grip via pressure sensors embedded in the palm and fingers. It is powered by a battery pack [74].

Extending the 'smart glove' concept in another direction, is the Human Machine Interface (HuMI) concept for teleoperation of external assets via hand motions. In this case, the glove is embedded with motion detection sensors sensitive to subtle motions of the hand (e.g., fingers). Signals from the sensors are transmitted wirelessly to mobile devices (e.g., drones) as controls for movement in three dimensions roll, pitch, yaw, and throttle. Similar embedded sensors in helmets could be used to control camera motions [75].

9.11 Transportation Services: Launch, Lunar Orbit and Landing

The availability of launch services to destinations in cislunar space will translate into a permanent and ultimately sustainable robotic infrastructure there, and human operations on at least a semi-permanent basis, analogous to visits to the ISS. NASA is currently planning on using the Artemis heavy lift launch vehicle to send the Orion capsule to the moon, but it is no longer 'the only game in town'. The SpaceX Falcon Heavy and ULA Vulcan have demonstrated the capability to send payloads to the Moon. NASA has been working with both SpaceX and Blue Origin to provide landers to rendezvous with Orion crews (orbiting the Moon) to take them to the lunar surface. The SpaceX approach is to use the Starship to launch its Human Landing System (HLS) into Earth orbit to be refueled there, so that it can boost itself into rectilinear lunar orbit, and from there take Orion crews from a lunar orbiting station to the lunar surface. Though little progress has been made so far, NASA is planning to build and maintain a station in an NRHO orbit (Gateway), the major components to be delivered by SpaceX Falcon Heavies. Presumably, with a minor change in architecture, the SpaceX HLS could rendezvous with a human crew in Earth orbit instead of lunar orbit and take them directly to the Moon.

Commercial launch service providers have already demonstrated the capability to get payloads successfully to the Moon, in particular SpaceX Falcon Heavy (Intuitive Machines Odysseus Lander), and Rocket Lab Electron/Proton (CAPSTONE orbiter). As part of the Lunar Initiative, NASA created the CLPS program to subsidize the development of landers by aerospace companies, and actually provided funding for Masten, Astrobotic, Intuitive

Machines, Draper Labs, and Firefly Aerospace. Intuitive Machines successfully performed the first automated landing in February of 2024, in previously unexplored and rugged polar terrain. Intuitive Machines is also developing its own compact mini-drones, as described above. Developing a commercially viable automated infrastructure to provide comprehensive characterization of the lunar environment will provide the basis for important advancement in lunar surface reconnaissance, through monitoring surface processes and identify resources for future use.

9.12 In Situ Manufacturing and Resource Extraction

For longer duration stays beyond Earth orbit, being able to print replacement parts in situ under minimal gravity conditions in order to refurbish and repair equipment will be an essential need. This has turned out to be an important consideration for the Earth-orbiting Space Station, to minimize the need to deliver additional payload to orbit. Currently, such zero gravity 3D printing devices, from Made in Space and more recently Braskem, are being developed and tested on the International Space Station [76, 77].

The technology for 3D polymer printing to replace or repair plastic parts has been evolving over the last decade. Currently, ISS is testing a device designed to accept plastics of a variety of compositions, sizes, and shapes as feedstock in order not only to fabricate but to reuse and recycle onboard material more efficiently [78]. More recently, Airbus has developed Metal3D, a capability to produce more resilient, load-bearing metal structures and tools. Unlike the polymer printers, the Metal3D uses wire, which acts independently of gravity, rather than powder [79]. SpaceCAL, a next generation 3D printer developed by Lawrence Livermore National Laboratory and being tested by Virgin Galactic, utilizes a new additive manufacturing technique to shape objects rapidly out of a viscous liquid made from a variety of materials, including silicones, glass composites, and biomaterials [80]. NASA is also funding the development of concepts for devices which excavate lunar regolith (Pilot Excavator) and extract icy volatiles (PRIME-1) [81, 82]. The lack of availability of lunar regolith simulant with morphologically and geotechnically analogous properties will mean that further refinements on such demonstration devices will be crucial work to be performed when during extended stays of the first crews visiting stations on a regular basis.

References

1. NASA Commercial Low Earth Orbit Development Program, Enabling US Leadership of In-Space manufacturing in Low-Earth Orbit for Terrestrial Markets - In Space for Earth, 2022. https://www.nasa.gov/humans-in-space/enabling-u-s-leadership-of-in-space-manufacturing-in-low-earth-orbit-for-terrestrial-markets-in-space-for-earth/. Visited 120124.
2. Pieters, C., J. Goswami, R. Clark, et al. Character and spatial distribution of OH/H2O on the Moon seen by M3 on Chandrayaan-1, Science, 326, 568-572, 2009.
3. Sunshine, J. M., T. Farnham, L. Feaga, et al Temporal and Spatial Variability of Lunar Hydration as Observed by the Deep Impact Spacecraft, Science, 326, 565-578, 2009.
4. Committee on the Scientific Context for Exploration of the Moon, Space Studies Board, National Research Council, 2007.
5. Cohen, B.A., Lunar Mission Priorities for the Decade 2023-2033, Lunar Exploration Analysis Group, 2023. https://ntrs.nasa.gov/api/citations/20205003736/downloads/Barbara%20Cohen_%20LEAG%20Mission%20Priorities%20White%20Paper.pdf
6. NASA Lunar Surface Innovations Initiative, Advancing transformative capabilities for lunar surface exploration across NASA's Space Technology Portfolio, 2024. https://www.nasa.gov/space-technology-mission-directorate/lunar-surface-innovation-initiative/. visited 010125
7. Fernholz, T., How Nokia and Axiom are Putting 4G on the Moon, Payload, August 27, 2024. https://payloadspace.com/how-nokia-and-axiom-are-putting-4g-on-the-moon/
8. Clark, P.E. and BIRCHES Lunar Ice Cube Team. Nature of and Lessons Learned from Lunar Ice Cube and the First Deep Space Cubesat 'Cluster', Proc. SPIE Optics+Photonics, Cubesats and Nanosats for Remote Sensing II, SPIE Digital Library, 107690G (14), 2018.
9. Clark, P.E. Cubesats in Cislunar Space, Annual AIAA/USU Conference on Smallsats, SSC18-V-03.
10. Clark, P.E. Compact instrumentation for experiments on the lunar surface. Proc. SPIE Optics+Photonics, Cubesats and Nanosats for Remote Sensing IV, SPIE Digital Library, https://doi.org/10.1117/12.2569812. 2020.
11. Clark, P.E., R. Staehle, D. Bugby Handheld, surface deployed or rover mounted instruments, Lunar Surface Science Workshops, 5030.pdf, 2020.
12. Bugby, D. and J. Rivera. Thermal Technology Advancements for Extended-Duration Lunar Operations, 52nd International Conference on Environment Systems, July 2023.
13. P.E. Clark, S.A. Curtis, M.L. Rilee, C.Y. Cheung, R. Wesenberg, J. Dorband, G. Brown. The next generation of tetrahedral rovers, in Space Technology and

Applications International Forum (STAIF-07), Ed. M.S. El-Genk, AIP Conf. Proc., 880, 711–718, 2007.
14. Williams, J-P, D. Paige, B. Greenhagen, E. Sefton-Nash. The global suface temperatures of the Moon as measured by the Diviner Lunar Radiometer Experiment. Icarus, 283, 300-325, 2017.
15. Frazier, S. NASA News-release, Three companies to help NASA advance solar array technology for Moon 2022. https://www.nasa.gov/news-release/three-companies-to-help-nasa-advance-solar-array-technology-for-moon/. visited 010125.
16. P.E. Clark, P.S. Millar, P.S. Yeh, S. Feng, D. Brigham, B. Beaman. Small Cold Temperature Instrument Packages, SPESIF011, edited by G. Patterson, AIP Conference Proceedings Physics Procedia, 20, 300–307, 2011.
17. Maki, G. and P-S Yeh, Radiation Tolerant Ultra Low Power CMOS Microelectronics: Technology Development Status, Semantic Scholar, ID: 18759505, 2003. https://www.semanticscholar.org/paper/Radiation-Tolerant-Ultra-Low-Power-CMOS-%3A-Status-sMaki-Yeh/ade61a5f8613d486325b6d4bb8efb4d1dc8774e0. visited 100124
18. Benz, H., J. Gambles, S. Whitaker, K. Hass, G. Maki P-S Yeh. Low power radiation tolerant VLSI for advanced spacecraft, IEEE Aerospace Conference Proceedings, 2003.
19. Mendelsohn, K. How to prevet another Europa Clipper Transistor Panic: 4 technologies that can radiation-harden future spacecraft electronics. IEEE Spectrum, January 2025. https://spectrum.ieee.org/europa-clipper. visited 010125
20. Somayajula, D., M. Crow An integrated dynamic voltage restorer-ultracapacitor design for improving power quality of the distribvution grid, IEEE Transactions on Sustainability, 6, 2, 616-624, https://doi.org/10.1109/TSTE.2015.2402221, 2015. https://ieeexplore.ieee.org/document/7063313. visited 100124.
21. Cortés-Peña, A. Y., T. D. Rolin, and C. W. Hill. A Novel Solid State Ultracapacitor. No. M17-6033. 2017.
22. Zhang, L., Shan, X., Bass, P. et al. Process and Microstructure to Achieve Ultra-high Dielectric Constant in Ceramic-Polymer Composites. Sci Rep 6, 35763. 2016.
23. Hofmann, D., L. Andersen, J. Koladziejska, S. Robert, J-P Borgonia, W. Johnson, K. Vecchio, A. Kennett. Optimizing buld metallic glasses for robust, highly wear-resistant gears. Advanced Engineering Materials, https://doi.org/10.1002/adem.2021600541, 2016.
24. JPL Robotics, NASA's Trio of Mini Rovers will team up to explore the Moon, 2003. https://www.jpl.nasa.gov/news/nasas-trio-of-mini-rovers-will-team-up-to-explore-the-moon. visited 100124.
25. Herridge, L. Commercial CubeRover Test shows how NASA investments mature space tech, NASA CLPS program, 2020. https://www.nasa.gov/missions/artemis/clps/commercial-cuberover-test-shows-how-nasa-investments-mature-space-tech/. visited 010125.

26. Luabeya, M., Commercial Moon Rovers under Test, NASA Artemis Program, 2024. https://www.nasa.gov/image-article/commercial-moon-rovers-under-test/. visited 010125.
27. NASA Artemis Program, Pressurized Rover, 2024. https://www.nasa.gov/extravehicular-activity-and-human-surface-mobility/pressurized-rover/. visited 100124.
28. Burlaka, O. Drones will help to map the Moon, Universe Magazine, March 22, 2024. https://universemagazine.com/en/drones-will-help-to-map-the-moon/#:~:text=Engineers%20and%20scientists%20strive%20to,to%20better%20plan%20future%20missions. visited 100124.
29. Intuitive Machines. Intuitive Machines and NASA finalize contract for extreme lunar mobility spacecraft, July 21, 2021. https://www.intuitivemachines.com/post/intuitive-machines-and-nasa-finalize-contract-for-extreme-lunar-mobility-spacecraft. visited 100124.
30. Catwell, MIT proposes exploring the Moon with a hovercraft, Element 14, 2022. https://community.element14.com/technologies/industrial-automation-space/b/blog/posts/mit-proposes-exploring-the-moon-with-a-hovercraft. visited 100124.
31. Evans, S., Hopping space robot tested in zero gravity, IOT World Today, April 12, 2024. https://www.iotworldtoday.com/robotics/hopping-space-robot-tested-in-zero-gravity. visited 100124.
32. Del Guercio, G. Science Today: Kangaroo Energy Efficiency-Just like a pogo stick. UPI.com. January 8, 1986. https://www.upi.com/Archives/1986/01/08/Science-Today-Kangaroo-Energy-Efficiency-Just-Like-A-Pogo-Stick/5720505544400/#:~:text=%2D%2D%20The%20red%20kangaroo%20of,much%20energy%20as%20a%20human. visited 010125.
33. Ackerman, E., Festo's Newest Robot is a Hopping Bionic Kangaroo. IEEE Spectrum April 2, 2014. https://spectrum.ieee.org/festo-newest-robot-is-a-hopping-bionic-kangaroo. visited 100124.
34. Kormushev, P., Kangaroo robot by Festo, Youtube, April 1, 2014. https://www.google.com/search?q=kangaroo+robot&rlz=1C5GCEM_enUS1073US1073&oq=kangaroo+robo&gs_lcrp=EgZjaHJvbWUqBwgAEAAYgAQyBwgAEAAYgAQyB-ggBEEUYOTIHCAIQABiABDIICAMQABgWGB4yCAgEEAAYFhgeMgoI-BRAAGAoYFhgeMggIBhAAGBYYHjIICAcQABgWGB4yCAgIEAAYFhgeMggI CRAAGBYYHggCCLACAQ&sourceid=chrome&ie=UTF-8#fpstate=ive&vld=cid:efd611ed,vid:_4luJ0ZSqy8,st:0
35. Clark, P.E., S.A. Curtis, M.L. Rilee. A new paradigm for Robotic Rovers, SPESIF2011, edited by G. Patterson, AIP Conference Proceedings, Physics Procedia, 20, 308–318, 2011.
36. Clark, P.E., S.A. Curtis, M.L. Rilee, W. Truszkowski, G. Marr, C.Y. Cheung, M. Rudisill, BEES for ANTS: Space Mission Applications for the Autonomous NanoTechnology Swarm, AIAA Intelligent Systems Technical Conference, Session 29-IS-13: ANTS, 2004.

37. Clark, P.E., M.L. Rilee, S.A. Curtis, C.Y. Cheung, G. Marr, W. Truszkowski, M. Rudisill. LARA: Near Term Reconfigurable Concepts and Components for lunar exploration and exploitation, IAC Proceedings, IAC-04-IAA.3.8.1.08, 2004.
38. Clark, P.E. M.L. Rilee, S.A. Curtis, C.Y. Cheung, G. Marr, W. Truszkowski, M. Rudisill. PAM: Biologically inspired engineering and exploration mission concept, components, and requirements for asteroid population survey, IAC Proceedings, IAC-04-Q5.07, 2004.
39. Clark, P.E., M.L. Rilee, S.A. Curtis, S.J. Bailin, S. Hall, E. Shaya, T. Speller, W. Truszkowski, B. Yatt, B. Zeigler, J. Nutaro, A. Powell, P. J. Costa. Reynolds. Frontier, Intelligent Decision Engine for Designing Stable Adaptable Complex Systems: An Overview, AIAA Space 2012, https://doi.org/10.2514/6.2012-5241, 2012.
40. NASA GSFC Tet Robotics Group, Mobility Extremes, 2010, https://www.youtube.com/@mobilityextremes
41. Curtis, S.A., M.L. Rilee, W. Truszkowski, C.Y. Cheung, P.E. Clark. Neural Basis Function Control of Super Micro Reconfigurable Technology (SMART) Nanostructures, AIAA Intelligent Systems Technical Conference, Session 29-IS-13: ANTS, 2004.
42. Patel, S., Impact Story: Terrain Relative Navigation, NASA STMD, 2022. https://www.nasa.gov/directorates/stmd/impact-story-terrain-relative-navigation/. visited 100124.
43. Clark, P.E., S. A. Curtis, M.L. Rilee, C.Y. Cheung, R. Wesenberg, J. Dorband, G. Brown. The next generation of tetrahedral rovers, in Space Technology and Applications International Forum (STAIF-07), Ed. M.S. El-Genk, AIP Conf. Proc., 880, 711–718, 2007.
44. Rilee, M.L., S.A. Curtis, C.Y. Cheung, P.E. Clark, W. Truszkowski. In-Space Autonomous Production Facility, AIAA Intelligent Systems Technical Conference, Session 29-IS-13: ANTS, 2004.
45. Schauer, K. and D. Baird, LunaNet: Empowering Artemi with Communications and Navigation Interoperability. NASA Communications, 2021. https://www.nasa.gov/humans-in-space/lunanet-empowering-artemis-with-communications-and-navigation-interoperability/. visited 100124.
46. JPL Technology, Beacon Monitor Operations, Deep Space 1 Technology, https://www.jpl.nasa.gov/nmp/ds1/tech/beacon.html. Visited 100124.
47. NASA Mission Operations and Communications Services, JPL Deepspace, 2009. https://deepspace.jpl.nasa.gov/files/NASA_MO&CS.pdf visited 100124.
48. Gardner, T., B. Cheethaam, J. Parker, A. Foresman, E. Kayser, M. Thompson, C. Ott, L. DeMoudt, M. Caudill, M. Bolliger, A. Kam, CAPSTONE: A summary of a highly successful mission in the cislunar environment, 17th Annual Small Satellite Conference, SSC23-1-04, 2023. https://digitalcommons.usu.edu/cgi/viewcontent.cgi?article=5595&context=smallsat. visited 100124.
49. Ono, M., B. Rothrock, Y. Iwashita, S. Higa, C. Timmeraju, S. Sahnoune, D. Qiu, T. Islam, A. Didier, C. Laporte, D. Atha, V. Sun, K. Otsu, M. Paton, O. Lamarre,

S. Daftry, R. Swan, A Stambouli, F. Chen, B. Shah, K. Stack, C. Mattmann, Machine learning for planetary rovers, in Machine Learning for Planetry Science, Elsevier Inc., 2022. https://doi.org/10.1016/B978.0-12-818721-0.00019-7. visited 100124.
50. Qiu, D., B. Rothrock, T. Islam, A. Didier, V. Sun, C. Mattmann, M. Ono, SCOTI: Science Captioning of Terrain Imagaes for data prioritization and local image search, Planetary and Space Science, 188, 104943, 2020.
51. Daftry, S., Z. Chen, Y. Cheng, S. Tepsuporn, S. Khattak, L. Matthies, LunarNav: Crater-based localization for long-range autonomous lunar rover navigation, IEEE Aerospace Conference, https://doi.org/10.1109/AERO55745.2023.10115640, 2023. https://ieeexplore.ieee.org/document/10115640. visited 100124.
52. Mathworks Help, BirdsEyeView, https://www.mathworks.com/help/driving/ref/birdseyeview.html. visitied 100124.
53. Swerdlow, A., R. Xu, B. Zhou, Street-View Image Generation from a Bird-s-Eye View Layout, arXiv:2301.04634v1 [cs.CV] 2023. https://arxiv.org/pdf/2301.04634v1. visited 100124.
54. Clark, P.E., S. Curtis, F. Minetto, J. Marshall, J. Nuth, SPARCLE: Electrostatic Dust Tool Proof of Concept. SPESIF-10, Ed. G. Patterson, AIP Conference Proceedings 1208, 549–556, 2010.
55. Abel, P., M. Anderson, E. Blom, C. Calle, P. Dunlap, P. Greenberg, D. Fisxher, S. Howard, K. Hurlbert, J. Jordan, D. Ludwiczak, E. Orndoff, F. Thomas, C Wohl, Lunar Dust Mitigation: A Guide and Referene, 1st Edition, NASA/TP-20220018746, 2021. https://ntrs.nasa.gov/api/citations/20220018746/downloads/TP-20220018746.pdf. visited 100124.
56. McKay, D.S., Heiken, G., Basu, A., Blanford, G., Simon, S., Reedy, R., French, B., Papike, J., "The Lunar Regolith", Lunar Sourcebook, Cambridge University Press, New York, N.Y., 285-356, 1991.
57. Carrier, W.D., G.R. Olhoeft, W. Mendell, "Physical Properties of the Lunar Surface," Lunar Sourcebook, Cambridge University Press, New York, N.Y., 475-568, 1991.
58. Gaier, J.R., The Effects of Lunar Dust on EVA Systems during the Apollo Missions, NASA TM-2005-213610, Glenn Research Center, NASA-TM-2005-213610, 2005.
59. Criswell, D.R., "Horizon-glow and the Motion of Lunar Dust," Photon and Particle Interactions with Surfaces in Space, D. Reidel Publ Co, Dordrecht, Holland, 545–556, 1973.
60. McCoy, J.E. and Criswell, D.R., "Evidence for a High Altitude Distribution of Lunar Dust," in the proceedings of the 5th Lun Sci Conference, Geochemica et Cosmochimica Acta, Supplement 5, 3, Houston, Texas, 2991-3005, 1974.
61. Berg, O.E., Wolf, H., Rhee, J.W., "Lunar Soil Movement Registered by the Apollo 17 Cosmic Dust Experiment," Interplanetary Dust and Zodiacal Light, Springer-Verlag, Berlin, Germany, 233-237, 1976.

62. Berg, O.E., "A Lunar Terminator Configuration," Earth and Planetary Science Letters, 39, 377-381, 1978.
63. Stubbs, T.J., Vondrak, R.R., Farrell, W.M., "A Dynamic Fountain Model for Lunar Dust," Adv Space Res, 37,1, 59-66, 2006.
64. Manka, R.H., "Plasma and Potential at the Lunar Surface," Photon and Particle Interactions with Surfaces in Space, D. Reidel Publ Co, Dordrecht, Holland, 347–361, 1973.
65. Cannon, K. C. Dreyer, G. Sowers, J. Schmit, T. Nguyen, K. Sanny, J. Schertz, Working with lunar surface materials: review and analysis of dust mitigation and regolith conveyance technologies, Acta Astronautica, 196, 259-274, https://doi.org/10.1016/j.actaastro.2022.04.037, 2022.
66. NASA Technology Transfer Program, Lotus Coatings, Materials and Coatings, GSC-TOPS-19, 2015. https://technology.nasa.gov/patent/GSC-TOPS-19#:~:text=The%20Lotus%20Coating%20is%20a,sticking%20to%20the%20coated%20surface. Visited 100124.
67. Zacny, K., B. Betts, M. Hedlund, P. Long, M. Gramlich, K. Tura, PlanetVac: Pneumatic regolith sampling system, IEEE Aerospace Conference, https://doi.org/10.1109/AERO.2014.6836409, 2014. https://ieeexplore.ieee.org/document/6836409. visited 100124.
68. Curtis, S.A., P.E. Clark, J. Marshall, F. Minetto, J. Marshall, J. Nuth, C. Calle, Observed Weak Electron Beam Discharge Driven Grain Acceleration/Accretion with Implications for Planet Formation. Earth, Moon, and Planets, 2010. http://www.springerlink.com/content/73rwm2v47521271g/?MUD=MP
69. Calle, C., A. Chen, S. Trigwell, Dust Particle removal by electrostatic and dielectrphoretic forces with applications to NASA exploration missions, Semantic Scholar, Corpus ID: 51807715, 2008.
70. Calle, C., M. Mazunder, C. Immer, C. Buhler, S. Clements, P. Lundeen, A. Chen, J. Mantovani, Electrodynamic dust shield for surface exploration activities on the Moon and Mars, 57th International Astronautical Congress, IAC-06-A5.2.07, 2006. https://arc.aiaa.org/doi/abs/10.2514/6.IAC-06-A5.2.07. visited 100124.
71. NASA Communications, Spacesuit for NASA's Artemis III Moon Surface Mission Debuts, 2023. https://www.nasa.gov/humans-in-space/spacesuit-for-nasas-artemis-iii-moon-surface-mission-debuts/. visited 100124.
72. Atkinson, N., NASA and Axiom Space do a partial reveal of the spacesuit that will be worn on the Moon, Universe Today, March 15, 2023. https://www.universetoday.com/160553/nasa-and-axiom-space-do-a-partial-reveal-of-the-spacesuit-that-will-be-worn-on-the-moon/. visited 100124.
73. Ugaide, V., Axiom spacw tests lunar spacesuit at NASA's Johnson Space Center, NASA Communications, January 29, 2024. https://www.nasa.gov/image-article/axiom-space-tests-lunar-spacesuit-at-nasas-johnson-space-center/. visited 100124.

74. NASA Spinoff, NASA's Robotic Glove Finds Commercial Handhold, NASA Technology Transfer Program, January 24, 2022. https://spinoff.nasa.gov/robotic-glove-finds-commercial-handhold. visited 100124.
75. Lee, P., and R. McDonald, An astronaut smart glove to explore the Moon Mars, and Beyond, SETI Isntitute, October 31, 2019. https://www.seti.org/press-release/astronaut-smart-glove-explore-moon-mars-and-beyond. visited 100124.
76. NASA Communications, International Space Station's 3D Priter, November 26, 2014. https://www.nasa.gov/image-article/international-space-stations-3-d-printer-2/. visited 100124.
77. Braskem News, Braskem and Made in Space to send plastics recycler to International Space Stations, November 4, 2019. https://www.braskem.com.br/news-detail/braskem-and-made-in-space-to-send-plastics-recycler-to-international-space-station#:~:text=The%20partnership%20between%20the%20two,to%20fabricate%20objects%20in%20space. Visited 100124.
78. NASA Missions, Full circle: NASA to demonstrate refabricator to recycle, reuse, repeat, International Space Station, August 28, 2017. https://www.nasa.gov/missions/station/full-circle-nasa-to-demonstrate-refabricator-to-recycle-reuse-repeat/. visited 100124.
79. Johnson, O., The world's first metal 3D printer for space developed by Airbus and AddUp to be tested on the ISS, TCT Magazine, January 31, 2024. https://www.tctmagazine.com/additive-manufacturing-3d-printing-news/metal-additive-manufacturing-news/airbus-sends-the-%E2%80%9Cworld%E2%80%99s-first%E2%80%9D-metal-3d-printer-for-space-developed-by-airbus-and-addup-to-be-tested-on-the-iss/. visited 100124
80. Ellery, M., Berkeley Researchers send 3D printer into space, UC Berkeley Engineering, July 2, 2024, https://engineering.berkeley.edu/news/2024/07/berkeley-researchers-send-3d-printer-into-space/. visited 100124.
81. Cawley, J., NASA Project takes off with New #D Lunar Simulation, NASA KSC, March 15, 2022. https://www.nasa.gov/centers-and-facilities/kennedy/nasa-project-takes-off-with-new-3d-lunar-simulation/. visited 100124.
82. NASA STMD, Polar Resource Ice Mining Experiment 1 (PRIME-1), Game Changing Development. https://www.nasa.gov/mission/polar-resources-ice-mining-experiment-1-prime-1/. visited 100124.

Afterword: Final Thoughts

To close, I want to focus on attributes, not necessarily thought of as related, that made the Apollo era Programs most successful: minimal to no micromanagement, along with maximum transparency, personal accountability, and proactive problem-solving.

These are attributes of 'the Bell Labs' model as that institution, during the decades it existed, was known for being the most successful generator of inventions and patents per capita that ever existed. The transistor is probably the best known example, but solar cells, communication satellite technology, video and cellular phones, fiber optic cables, quartz clocks, the UNIX computer, the discovery of Cosmic Microwave Background radiation, and prevention of UV degradation of most commonly used plastics all originated there. The ratio of scientists and engineers to managers was high. It was helpful of course that funding was long-term and stable and prestige and reputation were high. Short proposals originating from informal partnerships of staff scientists and engineers on research topics of interest were submitted every year. The rest of the time was spent doing research with funds made available as needed. As the excerpts from interviews in the opening chapter of this book illustrated, though the political climate was not as stable and the problem-solving was a bit more focused on 'aerospace' applications, many of these attributes existed and were encouraged among the engineers and scientists hired, as civil servants, contractors, or university grantees, and there were many more of them than managers. Once brought onboard, these folks worked with colleagues to tackle problems that no one had solved before, and, although time pressure was high, resources were readily made available, with paperwork to be handled by managers, to try approaches that looked promising. These efforts were discussed openly in the context of other groups efforts occurring simultaneously, not stove-piped. Hardware components or subsystems that met performance requirements had to be integrated with hardware from other developers, requiring that interfaces between components, systems, and subsystems be well documented and functional. This meant prioritizing control and standardization between interfaces, regardless of the developer or manufacturer, requiring complete transparency as one of the requirements of the contract. This approach was crucial to the success of NASA's Technology Transfer program, to expedite the availability of innovations for use in commercial revenue-generating ventures and thus part of the NASA charter. In recent years, even as requests for Non-Disclosure Agreements and Sensitive But Unclassified designations have proliferated, the actual transparency and insistence on standardized mechanical and electrical interfaces between complex systems has declined. This is exemplified by the fact two astronauts were stranded on International Space Station in mid 2024 due to incompatibilities between spacesuit connecter interfaces for different manufacturers. Starliner spacesuit connectors are designed for the Starliner vehicle only, and the Starliner capsule was unable to complete the docking for their trip home. For these astronauts to be returned to Earth, the Dragon capsule will need to bring up additional cargo of two dragon compatible spacesuits, which is not possible until mid 2025 at the earliest. The current DARPA LOGIC (Lunar Operating Guidelines for Interface Compatibility) effort to create a sustainable, shared, seamlessly integrated and interoperable lunar infrastructure from diverse assets should be used to address this issue.

For well over a decade now, the cubesat paradigm, an innovative and disruptive approach for space mission development, has proliferated. The initial goal was to create affordable opportunities for aerospace engineering students to build and test space hardware. However, the approach, which emphasizes rapid, grassroots development of compact, standardized platforms, has quickly become a primary approach for low Earth orbit, multi-platform, and technologically innovative (and thus risky) missions. The result has greatly increased demand for access to space, and thus economic viability and sustainability for space activities. This breakthrough resulted not only from the microminiaturization trend that began in the Apollo era, but also from a deliberate effort to leave behind the increasingly risk-averse micromanagement trend that has limited the breakthrough innovation in space in the intervening years. Clearly, this approach, with legacy from the Apollo era, should continue to be an important line for new business in the future.

Afterword: Final Thoughts

The *Springer-Praxis Space Exploration* series covers all aspects of human and robotic exploration, in Earth orbit and on the Moon and planets. The books tell behind-the-scenes stories of early and modern missions, both crewed and uncrewed, and cover all aspects of the space programs run by both leading and emerging spacefaring nations.

* * *

The books in this series are well illustrated with color figures and photographs. They are written in a style that space enthusiasts and historians, readers of popular magazines such as *Spaceflight* and readers of *Popular Mechanics* and *New Scientist* will find accessible.

Index

A

Advanced autonomy for dynamic architecture, 264–265
Advanced autonomy for locomotion, 261–264
ALSEP, 171–176
ALSEP, Environmental Experiments, 187–199
ALSEP, Interior Experiments, 176–187
ALSEP, Results Summary, 199–200
Amundsen, Antarctic Expeditions, 42–52
Antarctic expeditions, 42–52
Apollo, Cameras, 163–168
Apollo Current Online Resources, 1–5
Apollo Field Documentation Methods, 143–147
Apollo Field Simulations, 130–131
Apollo Field Tools, 155–168
Apollo Field Training, 108–130
Apollo Field Work Methodology Capture (on film), 133–138
Apollo Field Work Practices, 147–152
Apollo, Geological Toolkit, 159, 165–167
Apollo Lunar Roving Vehicle, 133–145
Apollo Lunar Surface Experiment Pack-age, *see* ALSEP
Apollo Lunar Surface Journal Excerpts, 143, 144, 147
Apollo Mission 11, 205–209
Apollo Mission 12, 205–209
Apollo Mission 14, 209–210
Apollo Mission 15, 211–214
Apollo Mission 16, 214–217
Apollo Mission 17, 218–221
Apollo Mission 18, 222–226
Apollo Mission 19, 222–226
Apollo Mission 20, 222–227
Apollo Missions, Cancelled (18, 19, 20), 222–227
Apollo Missions, J, 210–222
Apollo Missions, Pre-landing Crewed (7, 8, 9, 10), 87–96
Apollo, Sample Collection, 158–168
Apollo, Site Characterization, 160–168
Apollo Surface Mobility Options, 138–141
Apollo, Surface Sampling, 150, 155–158
Astronaut Gear, 273–274

Index

Astronaut Selection Criteria, 102–104, 106, 107, 109
Astronauts, Screened Women, 103
Astronauts Selected, 100–109
Astronaut Training, 111, 114–117, 125–130
Astronaut Training, Final Field Simulations, 131
Astronaut Training Sites, 1963–1968, 110–124
Astronaut Training Sites, 1969–1972, 114–117, 125–130

C

Cuff Check Lists for Apollo Missions, 144
Current Challenges, Communication Service, 266–269
Current Challenges, Documentation, 252
Current Challenges, Dust Mitigation, 269–273
Current Challenges, In Situ Manufacturing and Resource Extraction, 275
Current Challenges, In Situ Tools and Compact Package Networks, 250–251
Current Challenges, Navigation, 266–269
Current Challenges, Power, 252–254
Current Challenges, Space Suits, 273
Current Challenges, Thermal, 252–254
Current Challenges, Transportation Service, 274–275

E

El Baz, Farouk, interview, 5–8
Expeditions, Amundsen, 42–52
Expeditions, Antarctic, 42–52
Expeditions, Comparison to Apollo, 52–57

Expeditions, Lewis and Clark, 32–37
Expeditions, Louisiana Purchase, 32–37
Expeditions, Scott, Antarctic Expeditions, 42–52
Expeditions, Shackleton, Antarctic Expeditions, 42–52
Expeditions, Stanley, 37–41
Expeditions, Trans-Africa, 37–41

G

Geological Toolkit, Apollo, 158–168

H

Head, Jim, Interview, 13–21

I

Interviews, Farouk El Baz, 5–8
Interviews, Harrison Schmitt, 21–28
Interviews, Jim Head, 13–21
Interviews, Lee Silver, 9–13

L

Lunar Orbiter Program, 69–77
Lunar Surveyor Program, 77–86

M

Maps for Apollo missions, 147–152

P

Post-Apollo American Lunar Missions, 249
Post-Apollo Missions, proposed, 226–242
Post-Apollo Missions, proposed, region-al studies, 231, 235
Post-Apollo Missions, proposed, sortie studies, 230–234

R
Ranger Program, 62–75
Rovers, All Lunar to date, 255
Rovers, Apollo Lunar Roving Vehicle, 134, 140, 142
Rovers, Proposed and Advanced, 255–264

S
Schmitt, Harrison, Interview, 21–28
Scott, Antarctic Expeditions, 42–52

Shackleton, Antarctic Expeditions, 42–52
Silver, Lee, Interview, 9–13
Space Suits, 141
Space Suits, Apollo, 141
Space Suits, Current Challenges, 273
Stanley, Trans-Africa Expeditions, 37–41

T
Tetrahedral rover, 255

GPSR Compliance

The European Union's (EU) General Product Safety Regulation (GPSR) is a set of rules that requires consumer products to be safe and our obligations to ensure this.

If you have any concerns about our products, you can contact us on

ProductSafety@springernature.com

In case Publisher is established outside the EU, the EU authorized representative is:

Springer Nature Customer Service Center GmbH
Europaplatz 3
69115 Heidelberg, Germany

www.ingramcontent.com/pod-product-compliance
Lightning Source LLC
Chambersburg PA
CBHW070221311025
34791CB00028B/214

9 783032 017338